A Normal Distribution Course

T0175187

Jürgen Groß

A Normal
Distribution Course

PETER LANG

Frankfurt am Main · Berlin · Bern · Bruxelles · New York · Oxford · Wien

Bibliographic Information published by Die Deutsche Bibliothek
Die Deutsche Bibliothek lists this publication in the Deutsche Nationalbibliografie; detailed bibliographic data is available in the internet at <http://dnb.ddb.de>.

ISBN 3-631-52934-1
US-ISBN 0-8204-7348-0

© Peter Lang GmbH
Europäischer Verlag der Wissenschaften
Frankfurt am Main 2004
All rights reserved.

Printed in Germany 1 2 4 5 6 7

www.peterlang.de

Preface

I
know
of scarcely
anything so
apt to impress
the imagination
as the wonderful
form of cosmic or-
der expressed by the
"Law of Frequency of
Error." The law would
have been personified by
the Greeks and deified, if
they had known of it. It reigns
with serenity and in complete
self-effacement, amidst the wildest
confusion. The huger the mob, and
the greater the apparent anarchy, the
more perfect is its sway. It is the supreme
law of Unreason. Whenever a large sample
of chaotic elements are taken in hand and mar-
shaled in the order of their magnitude, an unsus-
pected and most beautiful form of regularity proves to
have been latent all along. *Sir Francis Galton*

Purpose and Scope

This monograph is intended as a treatise on the normal distribution in between a *probability theoretical* and a *data analytical* point of view. The normal distribution plays a central role in theoretical and applied statistics and there is a countless number of papers directly or indirectly referring to it. The *Handbook of the Normal Distribution* by Patel and Read [63] collects many results related to the normal distribution. We do neither intend to compete with nor to complete or extend it. Instead it is our purpose to discuss and illustrate some fundamental statistical methods in connection with the normal distribution. For this, proofs are omitted in general.

Albeit Galton's enthusiastic words quoted at the beginning[1], our text does not propagate the uncritical application of the normal distribution. Often, statistical methods developed under the assumption of normality turn out to be quite robust against deviations from normality. Nonetheless, analysis under certain forms of non-normality should not be suppressed, but, to the contrary, applied whenever necessary. If a variable tends to produce outliers being not in accordance with normality, then clearly this information should not be dismissed, but included in the analysis. If a distribution is skewed[2], some transformation to normality may be useful for certain purposes, but nonetheless knowledge about its skewness will be valuable in its own.

Hence, normality should not be regarded as a sovereign, but as a minister (in its original meaning as a servant) being valuable for specific purposes, but not to be followed uncritically.

Readers

This monograph is directed towards readers with at least basic knowledge in probability and statistics. It can be used for self-study purposes, conveying statistical principles behind inference based on data analysis. Emphasis is laid on the practicability of the presented methods, being mainly confined to the analysis of a single data set. Numerous graphics and examples with real and simulated data are included to illustrate the discussed topics. Mathematical derivations of the results are not given, since there are many books covering this field, some of which are cited in the references

---

[1] The quotation has been accredited to Galton by L.C. Tippett in: J.R. Newman, ed. (1956): The World of Mathematics. Volume Three. Simon and Schuster, New York. Page 1481.

[2] This is, for example, usually the case for scores on intelligence tests. See also the comments in: Devlin, B., Fienberg, S.E., Resnik, D.P., and Roeder, K. (1995): Wringing the Bell Curve: A Cautionary Tale About the Relationships Among Race, Genes, and IQ. Chance, **8**, pp. 27–36.

section. Eventually, it is hoped that even advanced readers will find the book a useful reference for properties and illustrations in relationship with the normal distribution.

Acknowledgements

For reading different earlier versions and providing a lot of helpful comments, I am grateful to Götz Trenkler and Barbara Bredner. Of course, responsibility for remaining errors is solely due to me.

Jürgen Groß

Contents

Chapter 1

Data Analysis

Usually, the starting point for a statistical analysis is a given set of n data points $x_1, x_2, \ldots, x_n$, which can be seen as measurements of a specific variable X of interest on n measurement units.

Then, the given data is supposed to reveal information about the variable X and especially about the probability law standing behind X. Such information may be used for statements like 'it is unlikely that X will take a value smaller than 0', 'the probability that X takes a value between 40 and 100 is about 0.8', 'the mean of a number of values of X is expected to be 66', 'the dispersion of X is quite strong, compared to...', etc.

In order to base such statements on a statistical analysis, one may start with a simple *description* of the data set and then continue with statistical *inference*, relying on basic assumptions about X and its observations (*the inference base*). The following three Sections 1.1, 1.2, and 1.3 give a short introduction into these concepts without laying claim on completeness.

1.1 Describing Data

When a data set is given, different tools are available for its description, some of which are discussed in the following. Reasonably, the application of such tools requires the use of a statistical software package.

1.1.1 Statistical Software Packages

Modern statistics cannot do without the use of a computer and a statistical software packages like SAS, SPSS, or S-PLUS (there are of course many others). Today, even personal computers are powerful enough to allow the handling of larger data sets as well as the application of rather involved procedures.

Unless otherwise stated, statistical computations and corresponding graphics in this text are done and generated with a variant of the S language called R. This statistical-computing environment has the appealing advantage of being freely available through the Internet under the General Public License (GPL). It can be downloaded from one of the CRAN (Comprehensive R Archive Network) sites, see the main site

`http://cran.r-project.org/`

See also [18, 45, 89] for how to do statistical analysis with R.

Although the methods described in our text are of general applicability and not explicitly related to a specific software, they are nonetheless oriented towards the R language.

1.1.2 Observations

As noted above, usually the starting point for statistical analysis is a given set of n observations.

Original Data and Summary Statistics

Let us consider the data given in Table 1.1 comprising $n = 100$ weights in kilogram (observations of the variable $X =$ 'weight') measured on female athletes (units of measurement) collected at the Australian Institute of Sport, see [15].

Clearly, from Table 1.1 it is difficult to see what is going on in the data set. To obtain a somewhat better impression, one may compute a summary statistics as shown in Table 1.2.

Min. and Max. give the minimum and maximum values occurring in the data set. First quartile (1st Qu.), median, and third quartile (3rd Qu.) divide the data set into fourths, i.e. about one fourth of the weights is smaller than 60.8, about one fourth of the weights is between 60.08 and

	Weight in kilogram									
1 – 10	78.90	74.40	69.10	74.90	64.60	63.70	75.20	62.30	66.50	62.90
11 – 20	96.30	75.50	63.00	80.50	71.30	70.50	73.20	68.70	80.50	72.90
21 – 30	74.50	75.40	69.50	66.40	79.70	73.60	78.70	75.00	49.80	67.20
31 – 40	66.00	74.30	78.10	79.50	78.50	59.90	63.00	66.30	60.70	72.90
41 – 50	67.90	67.50	74.10	68.20	68.80	75.30	67.40	70.00	74.00	51.90
51 – 60	74.10	74.30	77.80	66.90	83.80	82.90	64.10	68.85	64.80	59.00
61 – 70	72.10	75.60	71.40	69.70	63.90	55.10	60.00	58.00	64.70	87.50
71 – 80	78.90	83.90	82.80	74.40	94.80	49.20	61.90	53.60	63.70	52.80
81 – 90	65.20	50.90	57.30	60.00	60.10	52.50	59.70	57.30	59.60	71.50
91 – 100	69.70	56.10	61.10	47.40	56.00	45.80	47.80	43.80	37.80	45.10

Table 1.1: Weights of 100 female athletes, see [15]

Min.	1st Qu.	Median	Mean	3rd Qu.	Max.
37.80	60.08	68.05	67.34	74.43	96.30

Table 1.2: Summary statistics for the data from Table 1.1

68.05, about one fourth is between 68.05 and 74.43, and about one fourth is greater than 74.43. Mean stands for the usual arithmetic mean of the values.

Graphical Representation

Tables 1.1 and 1.2 give the data points and some corresponding characteristic numbers, which may be visualized by specific graphical methods.

The simplest way to illustrate a data set, is to plot the values on the x-axes (*strip chart*). See Fig. 1.1 (a), where in addition a *rug representation* is added, being essentially the same as a strip chart, but with the data points represented by small lines. Since there may exist data points which are identical or almost identical, one might get a better impression when the values are not plotted exactly on the x-axes but randomly moved above and below it to a certain amount. Such a plot is also called *jitter plot*. See

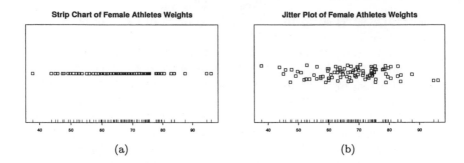

Fig. 1.1: Strip chart (a) and jitter plot (b) for the data from Table 1.1

Fig. 1.1 (b) for a jitter plot (with an additional rug plot) of the data from Table 1.1.

A more schematic plot related to the numbers in Table 1.2 is the so-called *box plot* or *box-and-whiskers plot*. Fig. 1.2 shows a box plot for the data from Table 1.1 compared to a corresponding jitter plot.

The box is drawn such that its left and right boundaries (the *hinges*[1]) are nearly first and third quartile, respectively, so that there is nearly 50% of the data within the range of the box. The ends of the two dotted lines (the *whiskers*) represent the smallest/largest value that falls within a distance of 1.5 times the box length from the lower/upper hinge. Data points beyond these are drawn separately. Very often a boxplot is displayed with x- and y-axes reversed, see Fig. 1.3 (a).

In Fig. 1.3 (a) we have also added a rug plot and a dotted horizontal line, representing the arithmetic mean of the data. Usually the mean is not included in a box plot. Box plots are especially useful for comparing the location and spread (dispersion) of several data sets. For example, in addition to the data from Table 1.1 we have also 105 male athletes weights available. A comparison of female and male weights via box plots is shown in Fig. 1.3 (b).

---

[1] Actually the *lower hinge* is the median of all those values being strictly smaller than the median, and the *upper hinge* is the median of all those values being strictly greater than the median.

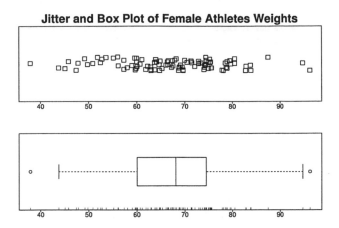

Fig. 1.2: Box plot (below) compared with corresponding jitter plot (above)

1.1.3 Frequency Distribution

The box plot gives an impression about the distribution of a given set of data points. Median, first and third quartile, minimal and maximal value can be seen as key features of *location* of a data set, while the *range* (maximal minus minimal value) and the *interquartile range* (IQR, third quartile minus first quartile, length of the box) can be seen as key features of *dispersion*. In addition, information about possible extreme values (*outliers*) is provided.

From the box plot in Fig. 1.3 (a), we may conclude that the data is rather symmetrically distributed about the median or the arithmetic mean. Both (median and mean) are almost identical, which is also an indication for symmetry.

As a further possibility, we may characterize the distribution of data in terms of frequencies.

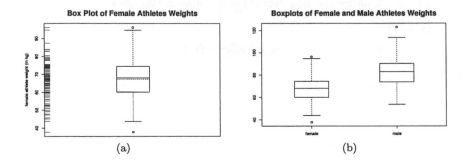

Fig. 1.3: Box plot (a) for the data from Table 1.1, compared (b) with boxplot for male athletes weights, see [15]

Computing Frequencies for Classes

When we have a large number of different observations of the variable X, then it is quite natural to specify classes and compute the absolute and relative frequency for each class. For the data from Table 1.1, classes of length 10 with corresponding frequencies are given in Table 1.3.

class	interval of weights	abs. freq.	rel. freq.
1	$(30, 40]$	1	0.01
2	$(40, 50]$	7	0.07
3	$(50, 60]$	17	0.17
4	$(60, 70]$	34	0.34
5	$(70, 80]$	32	0.32
6	$(80, 90]$	7	0.07
7	$(90, 100]$	2	0.02
Totals		100	1

Table 1.3: Weights of 100 female athletes

From Table 1.3 we can for example see that 66% of the female athletes have a weight greater than 60 and not greater than 80 kilograms.

Graphical Representation

A graphical representation of Table 1.3 can be constructed by drawing bars above each class interval with the height of the bar equal to the relative class frequency divided by the length of the class interval. See Fig. 1.4 for a *histogram* of the data from Table 1.1. Again, a rug representation of the data is added to the plot.

Note, that it is also possible to choose either relative or absolute frequency as the height of bars. As long as all class intervals have the same length, these choices will not give a different visual impression, but only yield a change in the scale of the y-axis.

Remark 1.1 A *true* histogram is build such that the *area* of each class bar is identical to the relative class frequency. Then the total area of the bars is identical to 1.

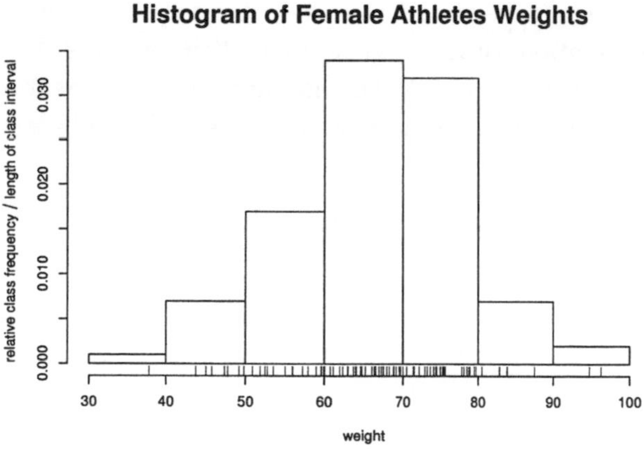

Fig. 1.4: Histogram for the data from Table 1.1

It is clear that the visual impression of a histogram also depends on the number of classes (length of class intervals) and the choice of the class intervals itself. Usually the classes are chosen to have equal length such that the data does not fall on class interval boundaries. For the number of classes there exist different proposals. The *Sturges' formula* suggests the smallest integer not less than $\log_2(n) + 1$ as the number of classes, where n is the number of observations. See also Sect. 3.2.1 for alternative class number proposals.

1.2   Inference Base

When we have observations $x_1, \ldots, x_n$ of a variable X of interest, then the description of the data can be the first step of analysis. When we make no further assumptions about the variable and its observations, it is more or less impossible to go to the next step of statistical inference.

The inference base we employ here will essentially consist of three assumptions:

- A value that X takes is related to the outcome of a random experiment (meaning that X is a *random variable*).
- It is possible to make some statements about the probability law under which observations of X are generated (*distribution* of X).
- A specific observation of X has no effect on or no relationship with any of the other observations (observations of X can be seen as values of a *random sample*).

Of course, these assumptions are not valid for all given data sets. Especially, the third assumption of independence of observations is not always satisfied.

1.2.1   Random Variable

We have noted above that the data at hand consists of observations on a variable X. This variable is assumed to be a *random variable*.

Essentially, a random variable X is a mapping assigning one (and only one) numerical value to each outcome of a random experiment. If we denote the set of all possible outcomes by Ω, then

$$X : \Omega \to \mathbb{R}$$
$$\omega \mapsto x \quad ,$$

and an actual value x of X depends on the actual result ω.

For example, if X is the weight of a person, then the random experiment will consist in the actual choice of a person ω, and the corresponding weight $x = X(\omega)$ is known, when the person has been chosen.

There are two important classes of random variables, namely the class of *discrete* and the class of *continuous* random variables.

Discrete Random Variable

A discrete random variable is defined as follows.

Definition 1.1 *A random variable X is called* discrete *if there exists a countable (finite or denumerable) set of values that X can take. These values are called (probability) mass points.*

If X is discrete, then of course also the set of values that X can take with *positive probability* is countable, since it is a subset of the set of mass points (usually both sets are identical).

As an example for a discrete random variable, consider the tossing of a single die and let X be the number of spots on the upper face. Then X can only take values from the set $\{1, 2, \ldots, 6\}$, being finite and thus countable.

A further example is the repeated tossing of a single coin, where X denotes the number of tails before the first head occurs. Then X can take values in $\{0, 1, 2, \ldots\}$, being denumerable and thus countable.

Continuous Random Variable

A major characteristic of a discrete random variable is, that it cannot take any value between two valid ones. On the other hand, when we consider a variable X like the weight of a person, then, in principle, this variable is allowed to take any value between two others, e.g. it is possible that any value between 40 and 100 kilograms may occur. Of course, in practice the weight is only measured to a certain degree of precision and thus it is not possible that any value will appear in our data set. But for the characterization of the essence of a continuous random variable the preciseness of measurement appears to be not relevant and thus, in principle, we have a nondenumerable set of possible values for X. In that case, the probability that a given value x will be taken by X should be infinitesimally small and therefore equal to zero.

Ad-hoc Definition 1.1 *A random variable X is called* continuous *if there does not exist a value that X can take with positive probability.*

From the above definition, a random variable X is continuous whenever $P(X = x) = 0$ for any given $x \in \mathbb{R}$. On the other hand, it is quite clear that the probability for X taking a value in a given interval will not necessarily be zero. For example the probability that the weight of a randomly chosen person is between 40 and 100 kilograms is definitely greater than zero. To meet this, a continuous random variable is usually defined via the existence of a specific function $f(\cdot)$.

Definition 1.2 *A random variable X is called* continuous *if there exists a function $f(x) \geq 0$ such that*

$$P(a \leq X \leq b) = \int_a^b f(x)\,\mathrm{d}x\,,$$

where $a \leq b$.

In other words, the probability that X takes a value within an interval $[a, b]$ is the area below some nonnegative function $f(x)$ considered in the interval $[a, b]$, see Fig. 1.5.

The motivation for introducing such a function $f(\cdot)$ will become clearer in the next section. At this stage, note that for a given value x we have

$$P(X = x) = \int_x^x f(t)\,\mathrm{d}t = 0\,,$$

meaning that the probability for X to take a specific value equals zero. Thus, Definition 1.2 is stronger than the previous ad-hoc definition and will henceforth be the valid definition for a continuous random variable.

1.2.2   Distribution of a Random Variable

So far, we have specified the difference between two kinds of random variables. When we are interested in the law of probability a random variable obeys (its *distribution*), we can see that there is a distinct difference between discrete and continuous random variables.

Continuous Random Variable X

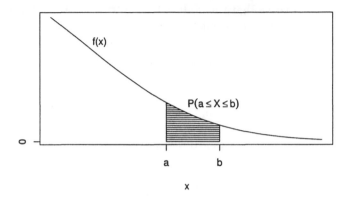

Fig. 1.5: Probability of $X \in [a, b]$ as area below a function $f(\cdot)$ in $[a, b]$

Discrete Distribution

Consider an experiment having only two results A and $\overline{A}$ with $P(A) = p$ and $P(\overline{A}) = 1 - p$, where p is unknown. Suppose that the same experiment is repeated n times, and each experiment is independent of the others. If we define $X =$ 'number of A', then we can conclude from combinatoric considerations that

$$P(X = x) = \binom{n}{x} p^x (1 - p)^{n-x}, \quad x = 0, 1, \ldots, n ,$$

see also Sect. 2.7.1. Here, if the experiment is carried out as described above, the probability of the mass points of X is known up to an unknown parameter p. Then, the only thing we have to find out is, how to obtain information about p from given data[2].

This describes a situation which *can* (but must not) occur in relationship with a discrete random variable: Once the experiment is performed,

---

[2]Presumably, we have a list of the results of the n experiments, and may then use the relative frequency of the occurrence of A as an estimate for p.

the probability of the mass points of X is known up to one or more unknown parameters, independent of actual observations of X.

Note that above we have introduced the term 'distribution' of X as
the 'law of probability' the random variable X obeys. We may now say,
that the distribution of a discrete random variable X is *characterized* by
its *probability mass function*, being defined below.

Definition 1.3 *Let X be a discrete random variable with mass points x_1,
$x_2, \ldots, x_n, \ldots$. Then the function*

$$f(x) = \begin{cases} \mathrm{P}(X = x_i) & \text{if } x = x_i,\ i = 1, 2, \ldots, n, \ldots \\ 0 & \text{otherwise} \end{cases}$$

is the probability mass function *(pmf) of X.*

Continuous Distribution

The situation is quite different when X is continuous. Then, usually we
do not know anything about the law of probability nature has given to a
specific X, e.g. we do not know how the weight of a person is distributed.

We can, nonetheless, make some assumptions about the distribution
when we assume that a large number of observations of X had been given.
In that case we could have divided the data into classes and have inspected
the relative class frequency, being an approximation of the true probability
for X taking a value within this class. When we have chosen small class
intervals, the corresponding histogram could have been approximated by
a smooth function $f(\cdot)$, see Fig. 1.6. This function is clearly nonnegative
(since the bar heights are nonnegative) and admits a total area of 1 (since
the total bars have an overall area of 1).

Definition 1.4 *Any function $f(\cdot)$ satisfying*

$$f(x) \geq 0\, \forall x \in \mathbb{R} \quad and \quad \int_{-\infty}^{\infty} f(x)\, \mathrm{d}x = 1$$

is called a probability density function *(pdf).*

A probability density function is linked to a specific random variable
X, if it is appropriate for computing probabilities related to X, i.e. if

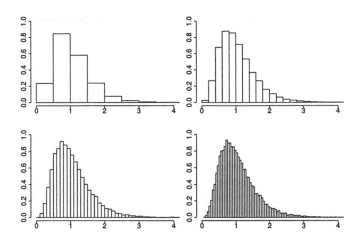

Fig. 1.6: Histograms with different number of classes for $n = 20000$ given data points

$$P(a \leq X \leq b) = \int_a^b f(x)\,dx\,,$$

for any $a \leq b$. In that case $f(\cdot)$ is called the pdf of X and it is identical to the function $f(\cdot)$ from Definition 1.2.

Remark 1.2 (The pdf is not necessarily continuous.) We have introduced the pdf as a smooth function replacing a histogram. This does, however, not mean that the pdf is necessarily a continuous function. As an example consider the pdf

$$f(x) = \frac{1}{3(b-a)}I_{[a,b]}(x) + \frac{2}{3(d-c)}I_{[c,d]}(x), \quad a < b < c < d\,,$$

where $I_B(x)$ denotes the indicator function, i.e. $I_B(x) = 1$ if $x \in B$ and $I_B(x) = 0$ if $x \notin B$ for some given subset $B \subset \mathbb{R}$. See Fig. 1.7 (a).

Remark 1.3 (The pdf does not give a probability.) For a continuous random variable X the pdf $f(x)$ for some given x does not equal $P(X = x)$.

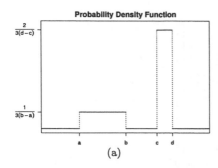
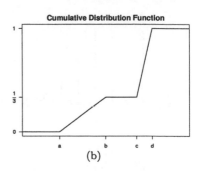

(a)   (b)

Fig. 1.7: Probability density function (a) and cumulative distribution function (b) of a continuous random variable

However, when we consider an interval $[x - \frac{1}{2}\Delta x, x + \frac{1}{2}\Delta x]$ for some small Δx, then the area of the rectangle with height $f(x)$ and width Δx may be seen as an approximation for

$$P(X \in [x - \tfrac{1}{2}\Delta x, x + \tfrac{1}{2}\Delta x]) = \int_{x-\frac{1}{2}\Delta x}^{x+\frac{1}{2}\Delta x} f(x)\,\mathrm{d}x \, ,$$

so that

$$f(x)\,\Delta x \approx P(X \in [x - \tfrac{1}{2}\Delta x, x + \tfrac{1}{2}\Delta x]) \, ,$$

i.e. the probability that X is in a *small* interval around x is approximately identical to the density $f(x)$ multiplied by the length Δx of the interval.

Cumulative Distribution Function

Strictly speaking, the pdf $f(x)$ of a continuous random variable X is not uniquely defined, since it is for example possible to change the function $f(x)$ at a finite number of points without changing the distribution of X. The simplest example is the pdf of the *uniform distribution* on an interval, which is usually defined by

$$f(x) = \frac{1}{b-a}I_{(a,b)}(x) \quad \text{or by} \quad f(x) = \frac{1}{b-a}I_{[a,b]}(x) \, .$$

Here, there are two different pdfs related to the open interval (a, b) or the closed interval $[a, b]$, but the distribution of X is the same in both cases.

There is, however, a function by which the distribution of any random variable X can be uniquely characterized.

Definition 1.5 *Let X be a random variable. Then the probability*

$$F(x) := P(X \leq x) \,,$$

considered as a function of $x \in \mathbb{R}$, is called the cumulative distribution function *(cdf) of X.*

Any cumulative distribution function can be characterized by three properties.

Theorem 1.1 *The function $F(x)$ with domain the real line and counter-domain $[0, 1]$ is a cdf if and only if the following three conditions hold:*

(a) $\lim_{x \to -\infty} F(x) = 0$ *and* $\lim_{x \to \infty} F(x) = 1$;
(b) $F(x_1) \leq F(x_2)$ *for $x_1 < x_2$, that is, $F(x)$ is a monotone, nondecreasing function;*
(c) $\lim_{0 < h \to 0} F(x + h) = F(x)$, *that is, $F(x)$ is continuous from the right.*

The cdf of a discrete random variable X is given by

$$F(x) = \sum_{x_i \leq x} P(X = x_i) \,.$$

It is necessarily a step function, where each step is at a position x_i for which $P(X = x_i) > 0$, compare Fig. 1.8 (a). The cdf of a continuous random variable is given by

$$F(x) = \int_{-\infty}^{x} f(t) \, \mathrm{d}t \,,$$

where $f(x)$ is the pdf of X. It is necessarily a continuous function but not necessarily *strictly* monotone, see also Fig. 1.7 (b). The cdf of the uniform distribution is

$$F(x) = \begin{cases} 0 & \text{for } x \leq a \\ \dfrac{x - a}{b - a} & \text{for } a < x < b \\ 1 & \text{for } b \leq x \end{cases} \,,$$

regardless of the choice of pdf

$$f(x) = \frac{1}{b-a} I_{(a,b)}(x) \quad \text{or} \quad f(x) = \frac{1}{b-a} I_{[a,b]}(x) \,.$$

We may even specify the pdf on $(a,b]$ or $[a,b)$ without changing the distribution and cdf.

In general, the cdf of a random variable must neither be a step function, nor continuous, but then the random variable X can neither be discrete nor continuous. Such random variables do of course exist.

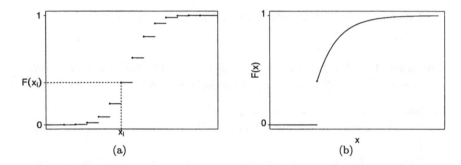

Fig. 1.8: Cumulative distribution function of a discrete random variable (a) and a neither discrete nor continuous random variable (b)

Expectation and Variance

When $f(\cdot)$ is the pmf or pdf of a discrete or continuous random variable X, then *expectation* and *variance* of X are two important measures to characterize the *central location* and the *dispersion* of $f(\cdot)$, respectively.

Expectation

Let us consider a discrete random variable X with only three mass points and corresponding probabilities given in Table 1.4.

Then, if we had observed a large number of values of X, we would have expected about 70% of them being identical to 1, about 20% being identical to 2 and about 10% being identical to 3. Now, when we had computed the

x	1	2	3
$P(X = x)$	0.7	0.2	0.1

Table 1.4: Mass points with corresponding probabilities of X

arithmetic mean of our values, then we would have expected it to be near to

$$1 \cdot 0.7 + 2 \cdot 0.2 + 3 \cdot 0.1 = 1.4 .$$

The value on the right-hand side is called the *expectation* of X. It gives the expected value of the arithmetic mean of a large (actually an infinite) number of values of X independent of actual observations of X. Compared to the arithmetic mean, being computed from a number of given values $x_1, \ldots, x_n$, the expectation is solely concluded from the pmf of X.

Definition 1.6 *Let X be a discrete random variable with mass points x_1, $x_2, \ldots, x_n, \ldots$ and pmf $f(\cdot)$. Then*

$$\mathrm{E}(X) = \sum_i x_i \, \mathrm{P}(X = x_i) = \sum_i x_i \, f(x_i) .$$

is called the expectation of X.

Hence, the expectation of a discrete random variable X is the average of its mass points weighted with the corresponding probabilities. For a continuous variable, the pdf $f(x)$ does *not* give the probability for X to take x. But from Remark 1.3, the product $f(x) \, \Delta x$ for some small Δx is approximately the probability for X to be in a small interval around x. Then weighting x with $f(x) \, \Delta x$ and considering the integral $\int x f(x) \, \mathrm{d}x$ instead of the sum $\sum x f(x) \, \Delta x$, gives the following definition for the expectation of a continuous random variable.

Definition 1.7 *Let X be a continuous random variable with pdf $f(\cdot)$. Then*

$$\mathrm{E}(X) = \int_{-\infty}^{\infty} x f(x) \, \mathrm{d}x$$

is called the expectation *of X.*

Remark 1.4 The expectation of X will also be denoted by μ_X or μ, the latter being used when it clear to what random variable the considered expectation belongs to.

Remark 1.5 The expectation of a random variable X can solely be concluded from its pmf or pdf $f(\cdot)$, and can thus also be seen as a characteristic value of $f(\cdot)$. Hence, we may call the expectation of X also the expectation of $f(\cdot)$.

Variance

The variance of a random variable X is the expected squared distance between X and $\mu = \mathrm{E}(X)$, when the squared distance is measured by $(X - \mu)^2$. When a number of outcomes of X can be observed, then a small variance implies that most observations will realize near μ, while in case of a greater variance not a few will realize quite far away from μ.

Definition 1.8 *Let X be a random variable mit* $\mathrm{E}(X) = \mu$. *Then*

$$\mathrm{Var}(X) = \mathrm{E}[(X - \mu)^2]$$

is called the variance *of X. The positive square root of* $\mathrm{Var}(X)$ *is called the* standard deviation *of X.*

A useful formula for the variance is

$$\mathrm{Var}(X) = \mathrm{E}(X^2) - \mu^2 \ ,$$

where $\mu = \mathrm{E}(X)$.

Remark 1.6 The variance of X will also be denoted by σ_X^2 or σ^2, the latter being used when it clear to what random variable the considered variance belongs to. The standard deviation will be denoted by σ_X or σ.

The standard deviation as a measure of the dispersion of $f(\cdot)$ is often preferred to the variance, since its unit of measurement is the same as that for the random variable X.

The Chebyshev Inequality

The Chebychev inequality gives a lower bound for the probability that X takes a value in an interval of length $2k\sigma_X$ around μ_X for any $k > 0$. More precisely, the Chebychev inequality says

$$P\left[X \in (\mu_X - k\sigma_X, \mu_X + k\sigma_X)\right] \geq 1 - \frac{1}{k^2}, \quad k > 0 .$$

This is true for any random variable X (discrete, continuous, or other) with finite variance. Choosing $k = 2$ gives

$$P\left[X \in (\mu_X - 2\sigma_X, \mu_X + 2\sigma_X)\right] \geq \frac{3}{4} ,$$

while $k = 3$ gives

$$P\left[X \in (\mu_X - 3\sigma_X, \mu_X + 3\sigma_X)\right] \geq \frac{8}{9} .$$

These inequalities may also serve as a justification for the use of μ_X as a measure of location and σ_X (or σ_X^2) as a measure of dispersion:

The probability that an actual value of X is within 2 standard deviations and 3 standard deviations of μ_X is at least 3/4 and 8/9, respectively. In other words, if we had a large number of values of X, then at least about 75% of the values would not be farther away from μ_X than $\mu_X - 2\sigma_X$ or $\mu_X + 2\sigma_X$. Moreover, at least about 88% of the values would not be farther away from μ_X than $\mu_X - 3\sigma_X$ or $\mu_X + 3\sigma_X$. Hence, μ_X can be seen as the center of the possible outcomes of X, although the values must not necessarily gather symmetrically about μ_X.

In addition, when σ_X is small, then the considered intervals are small and it is likely that an actual value of X is close to μ_X. Hence σ_X can be seen as a measure of dispersion in the sense that for a small σ_X most of the values of X will not be far away from their center.

However, the actual use of the Chebychev inequality is rather limited, since the true interval probabilities $P\left[X \in (\mu_X - k\sigma_X, \mu_X + k\sigma_X)\right]$ are often considerably greater than $1 - 1/k^2$. This also goes with the fact that the inequality makes no nontrivial statements for values $0 < k \leq 1$, which nonetheless would be of specific interest to assess the dispersion.

Example 1.1 Suppose we 'know' that the weight of person X (in kilogram) from a specific population has expectation $\mu = 75$ and standard deviation $\sigma = 14$. Then from the Chebychev inequality we can conclude that at least 88% of the persons from this population have a weight greater than 33 kilograms and smaller than 117 kilograms. But it may very well be the case that the true percentage between 33 and 117 kilograms is distinctly greater, e.g. 99%.   □

Joint Distribution

We have so far considered the distribution of a random variable X associated with a certain experiment. We may, however, also be interested in probabilities related to more than one random variable. For example, when we consider the weight X (in kilogram) and in addition the height Y (in cm) of a person, then we may be interested in probabilities such as

$$P(X \in [60, 90] \wedge Y \in [170, 190])\,,$$

etc. Thus we may consider the joint law of probability, i.e the *joint distribution*, of X and Y.

When we have random variables $X_1, \ldots, X_n$, then we may call

$$(X_1, \ldots, X_n)$$

an *n-variate* or *n-dimensional* random variable.

When random variables $X_1, \ldots, X_n$ are discrete, then analogously to the one dimensional case the *joint probability mass function* may be defined by

$$f(x_1, \ldots, x_n) = P\left[(X_1 = x_1) \wedge \cdots \wedge (X_n = x_n)\right]$$

if $(x_1, \ldots, x_n)$ is a (n-variate) mass point of $(X_1, \ldots X_n)$, and

$$f(x_1, \ldots, x_n) = 0$$

otherwise.

For the continuous case, a function $f(x_1, \ldots, x_n)$ is said to be a *joint probability density function* if

$$f(x_1, \ldots, x_n) \geq 0\,,$$

and

$$\int_{-\infty}^{\infty} \cdots \int_{-\infty}^{\infty} f(x_1, \ldots, x_n)\, dx_1 \ldots dx_n = 1\,.$$

It is linked to an n-variate random variable $(X_1, \ldots, X_n)$, if probabilities related to $(X_1, \ldots, X_n)$ can be computed from it. It that case $(X_1, \ldots, X_n)$ is called continuous. For example, if $n = 2$, $f(x_1, x_2) \geq 0$, and

$$P\left[(a_1 \leq X_1 \leq b_1) \wedge (a_2 \leq X_2 \leq b_2)\right] = \int_{a_2}^{b_2} \left[\int_{a_1}^{b_1} f(x_1, x_2) dx_1\right] dx_2 ,$$

then $f(x_1, x_2)$ is the joint pdf of the continuous random variable (X_1, X_2).

Independence

When we consider two random variables X and Y, then it may occur that the knowledge about the outcome of one variable, say X, changes the set of possible outcomes of the other variable Y, compared to the situation that nothing is known about the outcome of X.

For example, if X is the sum of the upturned faces of two tossed dices and Y is the absolut value of the difference of the upturned faces, then e.g. knowledge $X = 4$ restricts Y to be equal to 0 or to 2. This is of course different to the situation when nothing is known about X, in which case Y can take any value in $\{0, 1, 2, 3, 4, 5\}$. Hence the *conditional probability* of the event $[Y = 2]$ given the event $[X = 4]$ is $1/2$, while the *unconditional probability* of the event $[Y = 2]$ is $1/6$ and thus different. In that sense, Y depends on X. (It can be shown, that then also X depends on Y.)

Now, two discrete random variables X and Y are said to be *independent* of each other, if the conditional probabilities of one variable, given any value of the other variable, coincide with the corresponding unconditional probabilities. Hence, knowledge of the outcome of one variable has no influence on the probability for an outcome of the other variable. This is the case if and only if

$$f(x, y) = f_X(x) f_Y(y) ,$$

for all values x and y, meaning that the joint probability mass function of X and Y is the product of individual probability mass functions $f_X(x)$ and $f_Y(y)$ of X and Y, respectively. When X and Y are continuous random variables, then the probability density functions cannot directly be interpreted as probabilities, but the definition of independence remains the same for pdfs instead of pmfs. The notion of independence carries over to more than two random variables.

Definition 1.9 *Let $X_1, \ldots, X_n$ be random variables with joint pmf or pdf $f(x_1, \ldots, x_n)$. Then $X_1, \ldots, X_n$ are said to be independent, if*

$$f(x_1, \ldots, x_n) = \prod_{i=1}^{n} f_{X_i}(x_i) \, ,$$

for all $x_1, \ldots, x_k$, where $f_{X_i}(x_i)$ are the individual pmfs or pdfs of the random variables X_i, $i = 1, \ldots, n$.

Correlation

A special kind of dependence between two random variables X and Y occurs when

$$\varrho_{XY} := \frac{\mathrm{E}[(X - \mu_X)(Y - \mu_Y)]}{\sigma_X \sigma_Y}$$

is non-zero, in which case X and Y are called *correlated*. Correlatedness of X and Y can be seen as a kind of *linear* stochastic dependence between X and Y in the sense that if a number of pairs (x_i, y_i), $i = 1, \ldots, n$, of the two-dimensional random variable (X, Y) is observed and plotted in the xy plane, then the points scatter around an imagined straight line. The deviations of the points from this line will lessen with an increasing absolute value of ϱ_{XY}, where $|\varrho_{XY}| \leq 1$. Moreover, $|\varrho_{XY}| = 1$ if and only if there exist numbers a and $b \neq 0$ such that

$$\mathrm{P}(Y = a + bX) = 1 \, ,$$

i.e. Y is a linear transformation of X with probability 1.

Remark 1.7 If two random variables X and Y are correlated, i.e. $\varrho_{XY} \neq 0$, then X and Y are dependent. Equivalently, if two random variables X and Y are independent, then they are also uncorrelated, i.e. $\varrho_{XY} = 0$.

The converse of the above remark is not true. There can of course occur situations when X and Y are uncorrelated, but nonetheless dependent. Examples for this are given in Sect. 7.2.2.

1.2.3   Random Sample

When we have a number of observations $x_1, \ldots, x_n$ of a specific variable X, then we may assume that

- each observation has been generated under the same law of probability, and

- a certain value of one observation has no effect on or relationship with any of the other observations.

In that case the observations constitute what we may call a *random sample*. Of course, not every data set at hand satisfies these requirements, but often the experimental situation allows to proceed 'as if' (at least) they were fulfilled. On the other hand, if we have strong evidence that certain values of observations have an effect on others, then we cannot speak of a random sample in the above sense.

For a more formal approach, we assume that each observation x_i is the outcome of an individual random variable X_i with the same meaning as X. Then the two above assumptions can be written as

- each random variable X_i is distributed with the same pmf or pdf $f(x)$, and

- the random variables $X_1, \ldots, X_n$ are independent in the sense of Def. 1.9.

In that case we will call $X_1, \ldots, X_n$ the *random sample* and $x_1, \ldots, x_n$ the actual (n-dimensional) *sample value*.

Definition 1.10 *The random variables $X_1, \ldots, X_n$ are called a* random sample *of size n from the population $f(x)$, if they are independent and identically distributed with pmf or pdf $f(x)$.*

The joint pmf or pdf of a random sample $X_1, \ldots, X_n$ can be written as

$$f(x_1, \ldots, x_n) = \prod_{i=1}^{n} f(x_i) \,,$$

which may be used for calculating probabilities involving the sample.

1.3   Inference

For statistical inference, we will assume that our *data* consists of n observations $x_1, \ldots, x_n$ of a *random variable* X, and can be seen as the value of a *random sample* $X_1, \ldots, X_n$.

1.3.1   Estimation

When we have observed $x_1, \ldots, x_n$, then we may be interested in extracting information related to the distribution of X.

First, we may be interested in characteristic numbers of the distribution like *location* and *dispersion*. Hence, we may ask for some good replacement of the unknown expectation $\mu = \mathrm{E}(X)$ and variance $\sigma^2 = \mathrm{Var}(X) = \mathrm{E}[(X - \mu)^2]$. Such a replacement may be called an *estimate* (also: point estimate) of η, where η stands for μ, σ^2, or any other characteristic number (possibly multi-dimensional) of the distribution of X.

Since we are inclined to obtain an estimate from our observations, it will simply be a function $\gamma(x_1, \ldots, x_n)$ of them. The corresponding function $\gamma(X_1, \ldots, X_n)$ will be called an *estimator* for η. Any function of the sample being not necessarily meant as an estimator is simply called a *statistic*.

Definition 1.11 *If $X_1, \ldots, X_n$ is a random sample, then any function*

$$\gamma: \quad \mathbb{R}^n \quad \to \quad \mathbb{R}^m$$
$$(X_1, \ldots, X_n) \mapsto \gamma(X_1, \ldots, X_n)$$

with $\gamma(X_1, \ldots, X_n)$ not depending on unknown parameters, and being an m-dimensional random variable is called a statistic.

A statistic is considered as an estimator for the unknown m-dimensional η, if it can be considered as an appropriate replacement for η. What 'appropriate' actually means may depend on the researcher's goals under a given experimental situation. In general, a statistic should be considered as an estimator for a parameter of interest by reasons of both, common sense and statistical properties.

Definition 1.12 *If $X_1, \ldots, X_n$ is a random sample, then an m-dimensional statistic is called a* point estimator *for an unknown m-dimensional parametric function η, if it is considered as a replacement for η.*

Remark 1.8 (Estimator and Estimate) The actual outcome of an *estimator $G := \gamma(X_1, \ldots, X_n)$* is given by $g := \gamma(x_1, \ldots, x_n)$ and is then called the *estimate* of η.

A point estimate will not coincide with the parameter to be estimated, but a 'good' estimator will produce estimates which are likely to be near it.

To assess the performance of an estimator G, one may inspect its probability distribution[3], or, at least certain properties of its distributional behavior. Any estimator should reasonably have a relationship to the parameter to be estimated in the sense that the distribution of G depends on η.

For simplicity, we assume that η is a one-dimensional parameter in the following. Then G is a one-dimensional random variable, and key features of its distribution are expectation $\mathrm{E}(G)$ and variance $\mathrm{Var}(G)$. The difference

$$\mathrm{E}(G) - \eta$$

between the expectation of G and the unknown parameter η to be estimated is called the *bias* of G.

Ad-hoc Definition 1.2 *An estimator G is called a* good *estimator for η, if it has a small bias $\mathrm{E}(G) - \eta$ for all possible values of η and in addition a small variance $\mathrm{Var}(G)$.*

Thus, a 'good' estimator is supposed to produce estimates being likely to be near η. It is of course a further problem to assess what actually a 'small' bias and a 'small' variance means. Usually, we can only make some statements about the relative performance of a specific estimator in comparison with other estimators. Nonetheless, expectation and variance of an estimator are features of primary interest.

Unbiasedness

An often required property of an estimator is that it has zero bias. This means, an estimator G is called *unbiased* for some unknown parameter η, if

$$\mathrm{E}(G) = \eta$$

for all possible values of η. Unbiasedness of an estimator can be interpreted in the sense that if we had obtained not only one n-dimensional observation $(x_1, \ldots, x_n)$ of the random sample $(X_1, \ldots, X_n)$ but a large (actually

---

[3]Recall that G is a random variable.

infinite) number of such observations, and computed the corresponding estimates $g_1, g_2, \ldots$ for η, then the arithmetic mean of the estimates would (almost) coincide with η, see Fig. 1.9.

$$
\begin{array}{c}
(x_1, \ldots, x_n)_1 \to g_1 \\
\vdots \\
(x_1, \ldots, x_n)_h \to g_h
\end{array}
\qquad \to \frac{1}{h} \sum_{i=1}^{h} g_h \approx \eta
$$

Fig. 1.9: Arithmetic mean of h estimates from an unbiased estimator G for η

On the other hand, we usually have only one n-dimensional observation available, in which case we do not know whether the actual estimate is rather close or more far away from the parameter.

Hence, unbiasedness of an estimator alone cannot guarantee that this estimator is a good one. According to the above ad-hoc definition, we should in addition have some knowledge that its variance can be expected to be moderate, in which case it is likely that the estimate will not be far away from the expectation of G.

Consistency

The above ad-hoc definition claims an estimator G as 'good' for the unknown η, if it has small bias and small variance. Therefore, an optimal estimator has zero bias and zero variance. But this would mean that

$$
\mathrm{P}(G = \eta) = 1 \, ,
$$

so that the 'optimal' estimator 'knows' η with probability one. Clearly this is not a situation we assume to be confronted with in practice, so that an 'optimal' estimator in this sense will not exist.

Mean-Squared Error Consistency

On the other hand, we may very well expect the existence of such an 'optimal' estimator when the size n of the given sample approaches infinity.

Usually an estimator depending on the sample also depends on the sample size n. Intuitively, estimators should be better for greater sample sizes, since more information about the population distribution is delivered. When the sample size approaches infinity, it should be possible that the estimator approaches 'optimality', i.e.

$$\lim_{n\to\infty} [E(G_n) - \eta] = 0 \quad \text{and} \quad \lim_{n\to\infty} \text{Var}(G_n) = 0 \ ,$$

where $G \equiv G_n$ is indexed by n in order to emphasize the dependence on the sample size. As a matter of fact, estimators with this property frequently exist. Since

$$E[(G_n - \eta)^2] = [E(G_n) - \eta]^2 + \text{Var}(G_n) \ ,$$

it follows that the above two identities hold, if and only if

$$\lim_{n\to\infty} E[(G_n - \eta)^2] = 0 \ .$$

The expectation $\text{MSE}(G_n, \eta) := E[(G_n - \eta)^2]$ is also called the *mean squared error* (MSE) of G_n for η.

Definition 1.13 *The estimator G_n is called* mean squared error consistent *for η, if* $\lim_{n\to\infty} \text{MSE}(G_n, \eta) = 0$ *is satisfied.*

Mean-squared error consistency of an estimator G_n is a desirable property, since it means that G_n

- is *asymptotically unbiased* for η, i.e. $\lim_{n\to\infty} [E(G_n) - \eta] = 0$, and
- has asymptotically zero variance, i.e. $\lim_{n\to\infty} \text{Var}(G_n) = 0$.

Figure 1.10 illustrates mean-squared error consistency. Given an estimator G_n for a positive parameter η, each boxplot comprises the values of G_n obtained from 1000 samples of size n from the same underlying pdf. As can be seen, for any sample size n, the estimates do not exceed η, revealing that G_n is biased for η. Nonetheless, the estimates become closer to η for larger n. For a sample size of $n = 10000$, each of the 1000 estimates almost coincides with η, showing that bias and variance of G_n vanish for $n \to \infty$.

Weak Consistency

Often, consistency of an estimator is not considered with respect to vanishing bias and variance, but with respect to an arbitrary small interval

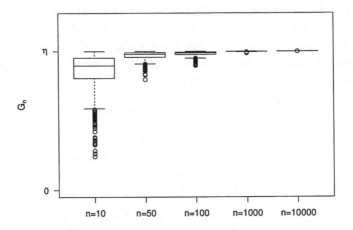

Fig. 1.10: Mean-squared error consistency of an estimator G_n for η

around η, in which the estimator takes its value with probability 1 for $n \to \infty$.

Definition 1.14 *The estimator G_n is called* weakly consistent *for η, if*

$$\lim_{n \to \infty} P[\eta - \varepsilon < G_n < \eta + \varepsilon] = 1$$

for every $\varepsilon > 0$.

It can be shown that a mean squared error consistent estimator G_n is also weakly consistent, but the converse does not hold. This is important because often the two definitions of consistency are mixed up. Although weak consistency can be interpreted as an asymptotic identity of estimator and parameter, it does not have the appealing meaning of an asymptotically vanishing bias and variance. One the other hand, in many situations, weakly consistent estimators are in fact also mean squared error consistent. This may be seen as the reason for the confusion: weak consistency is often proved via the stronger mean squared error consistency.

Nonetheless, if we can only diagnose weak consistency of an estimator, then we cannot even draw conclusion about its asymptotic unbiasedness.

The advantage of weak consistency lies in the fact that it can also be satisfied when mean squared error consistency is not, but still ensures a satisfactory asymptotic identity between estimator and parameter.

Example 1.2 Consider an estimator G_n for a parameter η, the latter being known to be not an integer. In addition, it is known that $P(G_n = \eta) = 1 - \frac{1}{\sqrt{n}}$ and $P(G_n = n) = \frac{1}{\sqrt{n}}$. Then for large n the probability $P(G_n = \eta)$ is almost 1, while the probability $P(G_n = n)$ is almost 0. As a matter of fact, G_n is weakly consistent for η. On the other hand, if we compute the expectation of G_n, then

$$E(G_n) = \eta(1 - \frac{1}{\sqrt{n}}) + n\frac{1}{\sqrt{n}} = \eta - \eta/\sqrt{n} + \sqrt{n} .$$

Obviously $\lim_{n\to\infty}[E(G_n) - \eta] = \infty$, so that G_n is not asymptotically unbiased for η, and thus not mean squared error consistent. □

The concept of weak consistency of an estimator G_n depending on the sample size n is related to the concept of *convergence in probability*.

Definition 1.15 *A sequence of random variables* $Y_1, Y_2, \ldots, Y_n, \ldots$ *is said to* converge in probability *to a random variable* Y, *if*

$$\lim_{n\to\infty} P[Y - \varepsilon < Y_n < Y + \varepsilon] = 1$$

for every $\varepsilon > 0$. *Convergence in probability is denoted by* $\operatorname{plim}_{n\to\infty} Y_n = Y$.

When we consider convergence in probability, then usually the random variables $Y_1, Y_2, \ldots, Y_n, \ldots$ are neither independent nor identically distributed, so that they are not a sample. Convergence in probability can also be considered to a fixed number v replacing a random variable Y. Hence, an estimator G_n (actually the sequence of estimators $G_1, G_2, \ldots, G_n, \ldots$) is consistent for the fixed parameter η, if and only if

$$\operatorname{plim}_{n\to\infty} G_n = \eta .$$

Closeness

Mean-squared error consistency of an estimator is an asymptotic property, guaranteeing the optimality of an estimator when the sample size n is infinite. For a fixed sample size, this optimality will of course only be attained to a certain degree, usually depending on whether the samples size n is larger or smaller. What actually 'large enough' for 'near optimality' is, will depend on the population distribution and the estimator itself. Since it is often costly and not always possible to increase the sample size as desired, it is also of interest to asses the performance of an estimator for a fixed sample size.[4]

It is quite obvious that the above mean squared error

$$\mathrm{MSE}(G, \eta) = \mathrm{E}[(G - \eta)^2]$$

of an estimator G for η can also be considered as a measure of goodness for a fixed sample size n. It gives the expected distance between the estimator and the parameter η, when the distance is measured by $(G - \eta)^2$. The smaller the distance, the more it is likely that the actual estimate is close to η.

Definition 1.16 *Let G_1 and G_2 be two estimators for a scalar parameter η. Then G_1 is called* uniformly *better* than G_2, if $\mathrm{MSE}(G_1, \eta) \leq \mathrm{MSE}(G_2, \eta)$ for all possible values η, and $\mathrm{MSE}(G_1, \eta) < \mathrm{MSE}(G_2, \eta)$ for at least one possible η.

It may of course be the case that the comparison of two estimators does not come up to the definite conclusion that one of the two is *uniformly* better than the other. It may happen that G_1 has smaller MSE for a subset of possible values η, but greater MSE for a different subset of possible values η. In such a case no definite conclusion can be drawn.

From the above definition of 'uniformly better', an estimator could be called 'best', if it uniformly better than any other estimator. Usually it is not possible to find an estimator having uniformly smallest MSE compared to any other possible estimator. Within certain subsets of estimators, however, it may be possible to identify a mean squared error optimal one. A very common approach is to confine to the set of unbiased estimators. For an unbiased estimator, the mean squared error coincides with its variance.

---

[4]Properties of an estimator for a fixed sample size are sometimes also called 'small sample' properties.

Definition 1.17 *An unbiased estimator G for a scalar parameter η is called* uniformly minimum variance unbiased estimator *(UMVUE), if for every unbiased estimator H of η the inequality $\mathrm{Var}(G) \leq \mathrm{Var}(H)$ holds true for all possible values of η.*

UMVUEs exist in many (but not all) situations. One might think that an UMVUE comes rather close to a 'good' estimator from the above ad-hoc definition, since it has zero bias and smallest variance. But this variance optimality is only within the set of unbiased estimators, and it may very well occur that there exists a biased estimator which has smaller variance and even uniformly smaller MSE than an UMVUE. In practice, however, the unbiasedness property is often regarded as a value in its own, and then an UMVUE is the best one can achieve.

Standard Error

From the Chebychev inequality[5] we know that

$$P\left[G \in (\mu_G - k\sigma_G, \mu_G + k\sigma_G)\right] \geq 1 - \frac{1}{k^2}, \quad k > 0\,,$$

where $\mu_G = \mathrm{E}(G)$ and $\sigma_G = \sqrt{\mathrm{Var}(G)}$. When the estimator G is known to be unbiased for the parameter η, then $\mu_G = \eta$, and for any given $k > 0$ the interval $(\eta - k\sigma_G, \eta + k\sigma_G)$ depends on the standard deviation σ_G of the estimator. The smaller σ_G, the more it is likely that our estimate will be near η. Hence, the value of σ_G is of some interest in its own, being often called the *standard error* of an actual estimate for η estimated by G.

One should, however, be aware that usually σ_G cannot directly be computed, since it will also depend on one or more unknown parameters. These must be estimated at first, so that the computed standard error is actually an estimate itself.

1.3.2 Sample Mean and Variance

In the previous subsection we have discussed general properties of point estimators. This has been motivated by our interest in the distribution of the random variable X, having generated the sample $X_1, \ldots, X_n$ with

---

[5]As we have already noted, the considered probability is often (distinctly) greater than the bound given by the inequality.

observations $x_1, \ldots, x_n$. Two characteristic numbers of the distribution of X are

$$\mu = \mathrm{E}(X) \quad \text{and} \quad \sigma^2 = \mathrm{Var}(X) = \mathrm{E}[(X - \mu)^2].$$

We will consider the *sample mean* and the *sample variance* as estimators for μ and σ^2.

Sample Mean

The sample mean is defined as

$$\overline{X} = \frac{1}{n} \sum_{i=1}^{n} X_i .$$

The actual estimate is then $\overline{x} = \frac{1}{n} \sum_{i=1}^{n} x_i$.

Theorem 1.2 *Let $X_1, \ldots, X_n$ be a sample from a distribution having expectation μ and finite variance σ^2. Then*

$$\mathrm{E}(\overline{X}) = \mu \quad \text{and} \quad \mathrm{Var}(\overline{X}) = \sigma^2/n$$

are the expectation and variance of the sample mean $\overline{X}$.

From the above theorem, the sample mean $\overline{X}$ is unbiased and mean squared error consistent for the population mean μ. The standard error of $\overline{X}$ is $\sigma/\sqrt{n}$, and is thus not known[6]. It is usually computed (more precisely: estimated) by $s/\sqrt{n}$, where s is the positive square root of the sample variance estimate s^2, see the discussion below.

Since $\overline{X} \equiv \overline{X}_n$ is mean squared error consistent for μ, it is also weakly consistent, i.e.

$$\operatorname*{plim}_{n \to \infty} \overline{X}_n = \mu .$$

This property is frequently named the *weak law of large numbers*.

The sample mean $\overline{X}$ can be used as an estimator for the unknown population mean μ when no further assumptions about the population distribution are made. It may, however, be inferior to other estimators in case that a specific distribution is assumed to stand behind the data, see Example 1.3, p. 36.

---

[6]Unless the population variance σ^2 is known.

Sample Variance

For $n > 1$ the sample variance is defined as

$$S^2 = \frac{1}{n-1} \sum_{i=1}^{n} (X_i - \overline{X})^2 .$$

The actual estimate is $s^2 = \frac{1}{n-1} \sum_{i=1}^{n} (x_i - \overline{x})^2$.

Theorem 1.3 *Let $X_1, \ldots, X_n$, $n > 1$, be a sample from a distribution having expectation μ and finite variance σ^2. Then*

$$\mathrm{E}(S^2) = \sigma^2 \quad \text{and} \quad \mathrm{Var}(S^2) = \frac{1}{n} \left(\mu_4 - \frac{n-3}{n-1} \sigma^4 \right) ,$$

where $\mu_4 = \mathrm{E}[(X - \mu)^4]$.

A proof for this result is given in [54, Theorem 2, Sect. VI.2]. From the theorem, the sample variance S^2 is unbiased and mean squared error consistent for the population variance σ^2.

Alternative Sample Variance

Instead of S^2 as an estimator, the statistic

$$D^2 = \frac{1}{n} \sum_{i=1}^{n} (X_i - \overline{X})^2$$

appears to be more intuitive and is sometimes also defined as the sample variance. From

$$\mathrm{E}(D^2) = \frac{n-1}{n} \sigma^2 \quad \text{and} \quad \mathrm{Var}(D^2) = \frac{(n-1)^2}{n^3} \left(\mu_4 - \frac{n-3}{n-1} \sigma^4 \right) ,$$

it can be seen that D^2 is not unbiased for σ^2. Nonetheless, D^2 is asymptotically unbiased and has asymptotically zero variance, so that it is mean squared error consistent for σ^2. For larger sample sizes n the difference between S^2 and D^2 is not significant in the sense that the actual estimates will be nearly identical.

Sample Standard Deviation

The *sample standard deviation* is defined as the positive square root $S = \sqrt{S^2}$ of the sample variance S^2. Usually S is employed as an estimator for the population standard deviation σ. As opposed to what one might guess at first thought, the sample standard deviation S is *not* unbiased for σ. This is due to the fact that the function $f(G)$ of an unbiased estimator G for η is not necessarily unbiased for $f(\eta)$. Even more, since the function $f : x \mapsto \sqrt{x}$ is concave for positive x, it follows from the Jensen inequality that

$$\mathrm{E}\left(\sqrt{S^2}\right) < \sqrt{\mathrm{E}\left(S^2\right)},$$

i.e. $\mathrm{E}(S) < \sigma$, as long as σ^2 and $\mathrm{Var}(S^2)$ are nonzero. On the other hand, the difference between $S \equiv S_n$ and σ asymptotically vanishes in the sense that

$$\plim_{n \to \infty} S_n = \sigma .$$

1.3.3   Maximum Likelihood Estimation

In the previous subsection, we have considered the estimation of the expectation μ and the variance σ^2 of the random variable X, standing behind the sample, when no further assumptions about the distribution of X are made. Quite often, however, we can assume that $X_1, \ldots, X_n$ is a random sample from a population $f(x; \theta)$. Here, $f(x; \theta)$ is a pmf or pdf, depending on a one- or multi-dimensional parameter θ from some parameter space Θ. Then, $\mathrm{E}(X) = \mu$ and/or $\mathrm{Var}(X) = \sigma^2$ will be functions $\tau(\theta)$ of the unknown parameter θ and our η from above actually becomes $\eta = \tau(\theta)$. Thus, in situations with data from $f(x, \theta)$, the primary interest lies in estimation of *parametric functions* $\tau(\theta)$. The most popular approach for this is the application of the maximum likelihood principle.

Likelihood Function

As we have already noted above, the joint density of the random variables $X_1, \ldots, X_n$ can be written as $f(x_1, \ldots, x_n) = \prod_{i=1}^{n} f(x_i; \theta)$. Let us now consider the joint density not as a function of possible observations $x_1, \ldots, x_n$, but as a function of the parameter θ for a particular value

$x_1, \ldots, x_n$. Actually, this particular value is taken as the observed sample value in the following.

Definition 1.18 *Let $x_1, \ldots, x_n$ be a sample value. Then the function*

$$L : \Theta \to \mathbb{R}$$
$$\theta \mapsto L(\theta; x_1, \ldots, x_n) = \prod_{i=1}^{n} f(x_i; \theta)$$

is called the likelihood function *of the sample.*

If $f(x; \theta)$ is a pmf, then

$$L(\theta; x_1, \ldots, x_n) = P(X_1 = x_1, \ldots, X_n = x_n)$$

the right-hand side also depending on the actual value of $\theta \in \Theta$. In this case, for any $\theta \in \Theta$ the likelihood function gives the probability for the occurrence of the (actually observed) sample $x_1, \ldots, x_n$. Hence, if

$$L(\theta_1; x_1, \ldots, x_n) > L(\theta_2; x_1, \ldots, x_n)$$

for two parameter values $\theta_1, \theta_2 \in \Theta$, then the probability for the occurrence of the sample value $x_1, \ldots, x_n$ is greater for $\theta = \theta_1$ than for $\theta = \theta_2$. From this, it appears reasonable to assume that the value θ_* for which the likelihood function becomes maximal should be near the true but unknown parameter θ in $f(x; \theta)$, having generated the observed sample $x_1, \ldots, x_n$.

The *likelihood principle* therefore tells us to choose that value θ_* as a replacement for the unknown θ which admits the highest probability for the occurrence of the already observed $x_1, \ldots, x_n$.

When $f(\cdot)$ is a pdf and not a pmf, then the likelihood function cannot directly be interpreted as the probability for the occurrence of $x_1, \ldots, x_n$. Nonetheless, if $f(x; \theta)$ is a pdf, then for some small $\varepsilon > 0$, the probability

$$P(x_1 - \tfrac{\varepsilon}{2} \le X_1 \le x_1 + \tfrac{\varepsilon}{2}, \ldots, x_n - \tfrac{\varepsilon}{2} \le X_n \le x_n + \tfrac{\varepsilon}{2})$$

for the occurrence of a value near of the already observed sample value $x_1, \ldots, x_n$ is

$$\prod_{i=1}^{n} P(x_i - \tfrac{\varepsilon}{2} \le X_i \le x_i + \tfrac{\varepsilon}{2}) \approx \prod_{i=1}^{n} \varepsilon f(x_i; \theta) = \varepsilon^n L(\theta; x_1, \ldots, x_n),$$

see Remark 1.3. Hence if

$$L(\theta_1; x_1, \ldots, x_n) > L(\theta_2; x_1, \ldots, x_n)$$

for two values $\theta_1, \theta_2 \in \Theta$, then also

$$\varepsilon^n L(\theta_1; x_1, \ldots, x_n) > \varepsilon^n L(\theta_2; x_1, \ldots, x_n) \,,$$

and the probability for the occurrence of a value near the observed sample value is greater for $\theta = \theta_1$ than for $\theta = \theta_2$. Again, it is reasonable to assume that the true but unknown parameter θ in $f(x; \theta)$, having generated the observed sample $x_1, \ldots, x_n$, is near to the value θ_* for which the likelihood function is maximized.

Definition 1.19 *The value* $\widehat{\theta} = \widehat{\theta}(x_1, \ldots, x_n)$ *is called* maximum likelihood estimate *for θ, if*

$$L(\widehat{\theta}; x_1, \ldots, x_n) = \max_{\theta \in \Theta} L(\theta; x_1, \ldots, x_n) \,.$$

The corresponding statistic $\widehat{\theta} = \widehat{\theta}(X_1, \ldots, X_n)$ *is the* maximum likelihood estimator *for θ.*

For some probability mass and density functions, maximum likelihood estimators can be derived as closed-form functions of the variables $X_1, \ldots, X_n$. For this, it is often convenient to consider not the likelihood itself, but the log-likelihood function

$$\ln[L(\theta; x_1, \ldots, x_n)] = \sum_{i=1}^{n} \ln[f(x_i; \theta)] \,.$$

As an alternative to maximizing the log-likelihood, one may also minimize the negative log-likelihood. Maximum likelihood estimators are known to have good asymptotic properties, see e.g. [54, Sect. VII 9].

Example 1.3 Consider observations $x_1, \ldots, x_n$ from a pdf

$$f(x, \mu) = \frac{1}{2\lambda} e^{-|x-\mu|/\lambda}, \quad -\infty < x < \infty \,,$$

where $\lambda = 1$ and μ is an unknown (but fixed) real number. This is a special case of the *double exponential distribution*, also called *Laplace distribution*, see e.g. [37, Chapter 24]. The expectation (as well as the median) of $f(x, \mu)$ equals μ. The likelihood function is given by

$$L(\mu; x_1, \ldots, x_n) = \frac{1}{2^n} e^{-\sum_{i=1}^n |x_i - \mu|} ,$$

and is maximized with respect to μ when $\sum_{i=1}^n |x_i - \mu|$ is minimized with respect to μ. It is well known, that the median of the x_i is the minimizer of the latter function. Hence, for n odd, $\widehat{\mu} = X_{((n+1)/2)}$ is the maximum likelihood estimator for μ, where $X_{(1)}, \ldots, X_{(n)}$ is the ordered sample. For n even, the median of the x_i is not unique, and as a matter of fact, the maximum likelihood estimator is not unique. One admissible solution is $\widehat{\mu} = \frac{1}{2}(X_{(n/2)} + X_{(n/2+1)})$, in which case our maximum likelihood estimator for μ becomes

$$\widehat{\mu} = \begin{cases} X_{((n+1)/2)} & \text{if } n \text{ is odd} \\ \frac{1}{2}(X_{(n/2)} + X_{(n/2+1)}) & \text{if } n \text{ is even} \end{cases} =: M ,$$

being also called the *sample median*. For the above pdf, the sample median M is known to be unbiased for μ, but it is *not* its UMVUE, see e.g. [37].

Nonetheless, the sample median M performs better than the sample mean $\overline{X}$, being also an unbiased estimator for μ here. From asymptotic considerations, see Theorem 2.11, p. 78, we may conclude that

$$\text{Var}(M) \approx \frac{1}{(2\sqrt{n} f(\mu_{0.5}))^2} ,$$

where $f(\cdot)$ is the pdf of the sample, and $\mu_{0.5}$ is the median of $f(\cdot)$. Since in the above case $\mu_{0.5} = \mu$ and $f(\mu) = 0.5$, it follows that the standard error of M is about $\sqrt{1/n}$. Since the variance of the above pdf is known to be $\sigma^2 = 2\lambda^2 = 2$, it follows that the standard error of $\overline{X}$ is $\sqrt{2/n}$, being greater than that of M. $\qquad \square$

In some cases it is not possible to find the maximum likelihood estimator in closed form. Then actual estimates can be obtained by some numerical optimization method using a computer.

Induced Likelihood Function

When we are interested in the estimation of some function $\tau(\theta) = \eta$, then we may consider the *induced likelihood function*

$$L^*(\eta; x_1, \ldots, x_n) = \sup_{\theta : \tau(\theta) = \eta} L((\theta; x_1, \ldots, x_n) \ .$$

The value $\widehat{\eta}$ for which $L^*(\eta; x_1, \ldots, x_n)$ is maximized will then be called the maximum likelihood estimator of $\eta = \tau(\theta)$.

Theorem 1.4 *[12, Theorem 7.2.10] If $\widehat{\theta}$ is the maximum likelihood esti- mator of θ, then for any function $\tau(\theta)$, the maximum likelihood estimator of $\tau(\theta)$ is $\tau(\widehat{\theta})$.*

The above theorem shows that it is reasonable to derive the maximum likelihood estimator $\widehat{\theta}$ for θ in $f(x, \theta)$ in a first step and then estimate $\tau(\theta)$ by $\tau(\widehat{\theta})$ in a second step. Note that θ can also be multi-dimensional.

1.3.4   Fitting Probability Density Functions

When X is a continuous random variable and $x_1, \ldots, x_n$ are given sample values of X, then there are basically two possible ways to fit a pdf $f(x)$ to the given data set:

- Consider the whole function as an unknown *parameter* and estimate it. This can be done via *kernel density estimators* with some given *kernel* and some given *bandwidth*.
- Take a parametric pdf $f(x; \theta)$ as a *model* for the distribution of X and estimate the unknown parameters in $f(x; \theta)$, in order to obtain the appropriate fit.

Kernel Density Estimation

Let $x_1, \cdots, x_n$ be the observations of a random sample with pdf $f(x)$. Then a *kernel density estimate* of $f(x)$ is a function of the form

$$\widehat{f}(x) = \frac{1}{nh} \sum_{i=1}^{n} K\left(\frac{x - x_i}{h}\right) \ ,$$

where $h > 0$, and $K(x)$ is a function satisfying $K(x) \geq 0$ for all x and $\int_{-\infty}^{\infty} K(x)\, dx = 1$. From this, it follows that

$$\widehat{f}(x) \geq 0 \quad \text{and} \quad \int_{-\infty}^{\infty} \widehat{f}(x)\, dx = 1 \ ,$$

showing that $\widehat{f}(x)$ has the properties of a pdf, see Def. 1.4. Different choices for the *kernel* $K(x)$ and the *bandwidth* h are possible.

The kernel is usually chosen as a pdf itself, the most common one being the *Gaussian kernel*

$$K(x) = \frac{1}{\sqrt{2\eta}} \exp(-x^2/2), \quad x \in \mathbb{R}.$$

Other kernels are e.g. the *Epanechnikov kernel*

$$K(x) = \begin{cases} \frac{3}{4}(1 - x^2) & \text{for } |x| < 1 \\ 0 & \text{otherwise} \end{cases},$$

the *triangular kernel*

$$K(x) = \begin{cases} 1 - |x| & \text{for } |x| < 1 \\ 0 & \text{otherwise} \end{cases},$$

or the *biweight kernel*

$$K(x) = \begin{cases} \frac{15}{16}(1 - x^2)^2 & \text{for } |x| < 1 \\ 0 & \text{otherwise} \end{cases}.$$

The choice of the bandwidth h is due to the user. When h is too large, the function $\widehat{f}(x)$ can become too smooth, thus hiding important features. On the other hand, when h is too small, many irregularities will veil the key features of the density. An often recommended bandwidth for the Gaussian kernel is

$$h = 1.06\sigma n^{-1/5},$$

where σ is the standard deviation of $f(x)$. Since the standard deviation is usually not known, it is required to use an estimate of σ instead. One may choose $\widehat{\sigma} = s = \sqrt{s^2}$ with

$$s^2 = \frac{1}{n-1} \sum_{i=1}^{n} (x_i - \overline{x})^2,$$

or

$$\widehat{\sigma} = \text{IQR}/1.34,$$

where IQR is interquartile range of the data. Often

$$\widehat{\sigma} = \min\{s, \text{IQR}/1.34\}$$

is proposed as an appropriate estimate for σ.

Example 1.4 Consider the $n = 100$ female atheletes weights $x_1, \ldots, x_n$ from Table 1.1. Then $s = 10.9155$ and $IQR/1.34 = 10.7090$, so that we may take $\widehat{\sigma} = \min\{s, IQR/1.34\} = 10.7090$ as our estimate for σ. When we apply the kernel density estimator for the Gaussian kernel, then the above rule of thumb for the bandwidth gives $\widehat{h} = 1.06\widehat{\sigma}n^{-1/5} = 4.5191$. Fig. 1.11 shows the resulting kernel density estimates $\widehat{f}(x)$ for the choices $h = 0.25\widehat{h}$, $h = \widehat{h}$, $h = 1.5\widehat{h}$, and $h = 4\widehat{h}$. $\qquad\qquad\qquad\square$

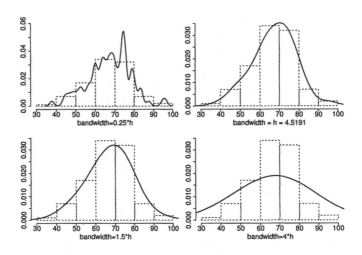

Fig. 1.11: Gaussian kernel density estimates using different choices of bandwidth for the female athletes weights data from Table 1.1

Parametric Density Estimation

Alternatively to the above procedure, we may assume that we know the class of density functions, our specific $f(x)$ comes from, so that

$$f(x) \in \{f(x, \theta) : \theta \in \Theta\}.$$

Here $f(x, \theta)$ denotes a pdf being completely known up to an unknown one- or multi-dimensional parameter θ from some parameter space Θ. Then, the remaining problem is to find an appropriate replacement for the unknown

θ, i.e. to find an appropriate point estimate for θ. In that case our density estimate becomes

$$\widehat{f}(x) = f(x, \widehat{\theta}).$$

The most common approach for finding point estimates $\widehat{\theta}$ is to apply the maximum likelihood principle described earlier.

Preciseness of Estimation and Goodness of Fit

The specification of the class of functions $\{f(x, \theta) : \theta \in \Theta\}$ by the user may also be seen as the imposition of some prior knowledge. The class of possible distributions is assumed to be known, but the corresponding parameters must be estimated from the data. The estimates can be said to be 'good' when they are close to the unknown parameters (preciseness of estimation), but also when they yield a good fit of the corresponding estimated density to the given data (goodness of fit).

In practice, these two properties do not always come together. When the sample value $x_1, \ldots, x_n$ perfectly reflects the distribution it comes from, then the best fit to the data should also produce very precise estimates, but usually the sample value reflects the underlying distribution only to a certain degree[7]. Then, if the chosen class $\{f(x, \theta) : \theta \in \Theta\}$ is flexible enough[8], the fit produced by the maximum likelihood approach may still look good, while the estimates can be considerably far away from the true parameters, so that a good fit does not in any case guarantee precise estimates. Of course, in the absence of further information, we are solely confined to the sample value for our analysis and cannot do better than the sample permits.

The AIC

The flexibility of a probability density function increases with its number of parameters, so that, in principle, the more parameter a pdf has, the higher becomes its potential for a good fit to the data. On the other hand,

---

[7]Especially for smaller sample sizes there may occur sample values for which a histogram of the data only remotely resembles the underlying distribution.

[8]Usually, this is the case for a multi-dimensional parameter θ, allowing for different shapes of the pdf $f(x, \theta)$.

as we have already noted in connection with kernel density estimation, if
the fit is 'too good', then it may even veil what is going on.

There are instances when a possible candidate pdf with a specific number of parameters is a special case of a pdf with more parameters. For
example the pdf of the *exponential distribution*

$$f(x) = \lambda \, e^{-\lambda x} I_{(0,\infty)}(x)$$

is a special case of the pdf of the *gamma distribution*

$$f(x) = \frac{\lambda^k}{\Gamma(k)} x^{k-1} e^{-\lambda x} I_{(0,\infty)}(x) \, ,$$

for the choice $k = 1$. When one of these pdfs seems to be appropriate, the
principle of parsimony tells us to fit the exponential distribution with only
one parameter, but possibly the data may tell us that the more flexible
gamma distribution with k different from 1 is more appropriate.

As a matter of fact, if parameter estimates are obtained by maximizing
the likelihood functions for both cases, the maximum value for the gamma
fit will never be smaller than the maximum value for the exponential fit,
i.e.

$$L(\widehat{\lambda}^{(2)}, \widehat{k}^{(2)}; x_1, \ldots, x_n) \geq L(\widehat{\lambda}^{(1)}, k \equiv 1; x_1, \ldots, x_n) \, ,$$

where $\widehat{\lambda}^{(2)}$ and $\widehat{k}^{(2)}$ are the maximum likelihood estimates under the
gamma distribution model, while $\widehat{\lambda}^{(1)}$ is the maximum likelihood estimate
under the exponential distribution model. Thus, the maximum likelihood
principle favors models with more parameters, and in order to take the
principle of parsimony into account some adjustment with respect to the
number of parameters is required.

A possible aid for deciding which of the two pdfs to choose can be the
so-called AIC (An Information Criterion) due to Akaike [3]. It is given as

$$\text{AIC} = -2 \left(\ln \left[L(\widehat{\theta}; x_1, \ldots, x_n) \right] - \dim(\theta) \right) \, ,$$

where $L(\widehat{\theta}; x_1, \ldots, x_n)$ is the actual likelihood computed from the observations $x_1, \ldots, x_n$ and the estimate $\widehat{\theta}$ of the (more-dimensional) parameter
θ, and the actual number $\dim(\theta)$ of elements of θ. When we have to choose
between two pdfs, one being a special case of the other, then we may compute the AIC for both cases, and choose the pdf with *smaller* AIC. This

criterion may also be applied when the actual estimates are not derived from the likelihood principle. The procedure does, however, not guarantee that the 'correct' distribution (in whatever sense) can be found. For example, the exponential distribution is also a special case of the *Weibull distribution* with pdf

$$f(x) = \lambda b x^{b-1} e^{-\lambda x^b} I_{(0,\infty)}(x)$$

for the choice $b = 1$, but the Weibull and the gamma distributions are not special cases of each other. The decision between the latter two distributions should not be based on AIC, since the corresponding likelihood functions are not directly comparable to each other.

An alternative to the AIC is the BIC (Bayesian Information Criterion), defined as

$$\text{BIC} = -2 \ln \left[L(\widehat{\theta}; x_1, \ldots, x_n) \right] + \ln(n) \dim(\theta) .$$

For sample sizes greater than 7 it penalizes models with greater number of parameters stronger than AIC.

1.4 A Summary Illustration

In this chapter we have outlined possible ways from data analysis to inference, allowing conclusions about the distribution of a random variable of interest. Fig. 1.12 aims to give a summary illustration for this, where of course only a very limited part of the discussed concepts can be taken into account. Here, the considered random variable is assumed to be continuous, the histogram is seem as an estimator for the pdf, and the characteristic of interest is the expectation $E(X)$.

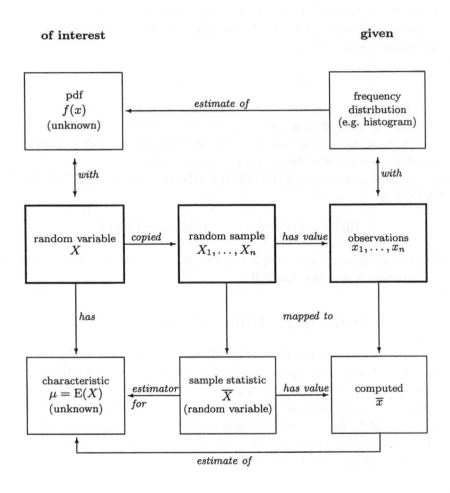

Fig. 1.12: Estimation of distributional properties of continuous X

Chapter 2

The Normal Distribution

The normal distribution plays a central role in techniques and methods of applied statistics. A sound historical review of its development is given in [63], see also [36] for historical remarks. As revealed by K. Pearson in 1924 [66], the normal distribution can be traced back to ABRAHAM DE MOIVRE (1667–1754), who derived it in 1733 as an approximation to the probability for sums of binomially distributed quantities, cf. [52] and [53, pp. 243–250]. In 1774, PIERRE SIMON LAPLACE (1749–1837) obtained the normal distribution as an approximation to the hypergeometric distribution [41]. Later his important statistical work often referred to properties in connection with the normal distribution.

The normal distribution as a law of error related to the method of least squares has been used by CARL FRIEDRICH GAUSS (1777–1855) in 1809

[27]. Although Gauss himself later did not rate the normal law as very important, his work helped to popularize its use in the following. Often, the normal distribution is also called the *Gaussian distribution*. The former German 10 DM banknote[1](used from 1991 to 2001), showed a portrait of Gauss together with a graph of the normal density function. The term 'normal law' has been used by FRANCIS GALTON (1822–1911) in 1877 and further be popularized by the work of KARL PEARSON (1857–1936).

2.1 Definition of the Normal Distribution

A random variable X is said to follow a *normal distribution*, if it has pdf

$$f(x) = \frac{1}{\sigma\sqrt{2\pi}}\, e^{-\frac{1}{2}(x-\mu)^2/\sigma^2}, \quad -\infty < x < \infty ,$$

where $-\infty < \mu < \infty$ and $0 < \sigma$ are parameters. Usually, σ^2 instead of σ is defined as the second parameter and the normal distribution is denoted by

$$N(\mu, \sigma^2) .$$

In connection with this denotation, the second parameter will always be σ^2, but in general we will refer to σ as well as σ^2 as the second parameter.

The normal pdf $f(x)$ is symmetric about $x = \mu$, having the often cited *bell shape*. It is also known as the *bell curve*. The pdf is unimodal and has inflection points at $\mu - \sigma$ and $\mu + \sigma$.

2.1.1 Location and Scale Parameter

The two parameters μ and σ determine the shape of the function in completely different directions. The parameter μ is solely responsible for the x-axis, i.e. it translates the function $f(x)$ to the location $x = \mu$ as its center. It is therefore also called the *location parameter* of the normal distribution. The parameter σ is solely responsible for the y-axis, i.e. it makes the function $f(x)$ flatter and more extensive around μ or higher and more concentrated around μ, see Fig. 2.2. Therefore, σ is also called the *scale parameter* of the normal distribution.

---

[1]The picture of the banknote is taken from the web site of the Deutsche Bundesbank: http://www.bundesbank.de/bargeld/euro_dm_banknoten.php

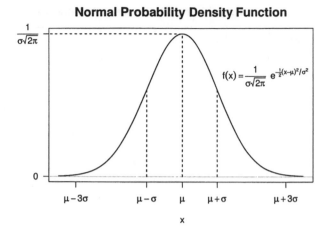

Fig. 2.1: Probability density function of the normal distribution

For a formalization, let $\phi(x)$ denote the function $f(x)$ for the choices $\mu = 0$ and $\sigma = 1$, i.e.

$$\phi(x) = \frac{1}{\sqrt{2\pi}} e^{-\frac{1}{2}x^2} .$$

Then the function $f(x)$ can be written as

$$f(x) = (1/\sigma)\phi[(x - \mu)/\sigma] ,$$

so that the family $\{f(x) : -\infty < \mu < \infty, 0 < \sigma\}$ is a *location-scale family with standard pdf* $\phi(x)$ according to the following definition.

Definition 2.1 [12, Definition 3.5.5] *Let* $\eta(x)$ *be any pdf. Then the family of pdfs*

$$\{h(x) = (1/b)\eta[(x - a)/b] : -\infty < a < \infty, 0 < b\}$$

is called the location-scale family with standard pdf $\eta(x)$*; a is called the* location parameter *and b is called the* scale parameter.

Since μ is a pure location parameter and σ is a pure scale parameter, and both parameters entirely specify the normal distribution, a linear

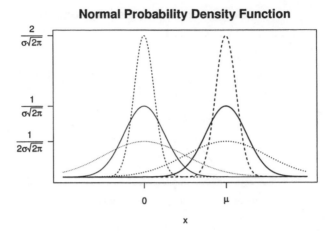

Fig. 2.2: Probability density function of the $N(\mu, \sigma^2)$ distribution (solid line), compared to the pdfs of $N(\mu, (\sigma/2)^2)$ and $N(\mu, (2\sigma)^2)$ distributions (dotted lines); the non-bold lines corresponding to the case $\mu = 0$

transformation of a normally distributed random variable X does not alter the type of distribution, but only location and scale. As a matter of fact, we have the following result.

Theorem 2.1 *Let X follow a $N(\mu, \sigma^2)$ distribution. Then for any scalars a and $b > 0$, the random variable $a \pm bX$ follows a $N(a \pm b\mu, b^2\sigma^2)$ distribution.*

2.1.2 Expectation and Variance

The parameters μ and σ are not only location and scale parameters, they are also identical to the expectation and standard deviation of a normally distributed random variable. As a matter of fact, it is a straightforward exercise to show that

$$\mathrm{E}(X) = \int_{-\infty}^{\infty} x \, \frac{1}{\sigma\sqrt{2\pi}} \mathrm{e}^{-\frac{1}{2}(x-\mu)^2/\sigma^2} \, \mathrm{d}x = \mu$$

and

$$E(X^2) = \int_{-\infty}^{\infty} x^2 \frac{1}{\sigma\sqrt{2\pi}} e^{-\frac{1}{2}(x-\mu)^2/\sigma^2} \, dx = \mu^2 + \sigma^2$$

so that $\mathrm{Var}(X) = E(X^2) - [E(X)]^2 = \sigma^2$. Hence, if we know that a variable is normally distributed, and we know expectation and variance of this variable, then its distribution is completely known. In other words, the normal distribution is entirely specified by its first two moments.

2.1.3 Alternative Parametrizations

The above parametrization is not the only possible one, although it is commonly used, and is also our choice of parametrization in the following.

Nonetheless, we may also say that a random variable has a normal distribution, if, for example, its pdf is given by

$$g(x) = \tfrac{1}{\lambda} e^{-\pi(x-\mu)^2/\lambda^2} \, ,$$

where $-\infty < \mu < \infty$ and $0 < \lambda < \infty$ are parameters[2]. By letting

$$\gamma(x) = e^{-\pi x^2} \, ,$$

it follows that $g(x) = (1/\lambda)\gamma[(x - \mu)/\lambda]$, so that

$$\{g(x) : -\infty < \mu < \infty, 0 < \lambda\}$$

is a *location-scale family with standard pdf* $\gamma(x)$ and location and scale parameters μ and λ. Here, the location parameter μ is again the expectation of X, but the scale parameter λ is not the standard deviation of X, the latter still being $\sigma = \lambda/\sqrt{2\pi}$.

Remark 2.1 In general, a location parameter does not necessarily coincide with the expectation and a scale parameter does not necessarily coincide with the standard deviation.

---

[2]The function $g(x)$ characterizes the same set of distributions as $f(x)$ in a different parametrization. We obtain $g(x)$ from $f(x)$ by putting $\sigma = \lambda/\sqrt{2\pi}$ and vice versa $f(x)$ from $g(x)$ by putting $\lambda = \sigma\sqrt{2\pi}$.

2.2   Standard Normal Distribution

As we have seen above, there are different possibilities for the definition of a *standard pdf* with respect to a location-scale family, depending on the parametrization we choose. There is, however, an alternative and often used notion of *standardization*. Namely, when we consider a random variable X with expectation $E(X) = \mu$ and variance $Var(X) = \sigma^2$, then the random variable

$$Z = \frac{X - \mu}{\sigma}$$

is called standardized in the sense that $E(Z) = 0$ and $Var(Z) = 1$.

2.2.1   The Function $\phi(x)$

When X is normally distributed with $E(X) = \mu$ and variance $Var(X) = \sigma^2$, then from Theorem 2.1 the random variable $Z = (X - \mu)/\sigma$ is $N(0,1)$ distributed and thus has pdf $\phi(x)$. Hence $\phi(x)$ is not only the standard pdf with respect to a specific parametrization, but it is also the pdf of a standardized normally distributed random variable. This is the reason why the distribution associated with the pdf

$$\phi(x) = \frac{1}{\sqrt{2\pi}} e^{-\frac{1}{2}x^2}$$

is called the *standard normal distribution*. We have already noted above, that the pdf $f(x)$ of $N(\mu, \sigma^2)$ is related to $\phi(x)$ via the equation

$$f(x) = (1/\sigma)\phi[(x - \mu)/\sigma] .$$

The relationship between the pdfs $f(x)$ and $\phi(x)$ has a counterpart in the relationship between the corresponding random variables, as stated in the following theorem, see [12, Theorem 3.5.6].

Theorem 2.2 *Let $\eta(\cdot)$ be any pdf, and let a and $b > 0$ be any scalars. Then X is a random variable with pdf $h(x) = (1/b)\eta[(x-a)/b]$ if and only if there exists a random variable Z with pdf $\eta(x)$ and $X = a + bZ$.*

The above theorem is valid for any pdf $\eta(\cdot)$. When applied to the standard normal pdf it shows that X has a $N(\mu, \sigma^2)$ distribution if and only if X can be written as $X = \mu + \sigma Z$, where Z has a standard normal distribution. In view of our previous remarks, this is of course not very surprising, but the above theorem is of value in its own.

2.2.2 The Function $\Phi(x)$

The cumulative distribution function of the standard normal distribution is

$$\Phi(x) = \int_{-\infty}^{x} \phi(t)\,dt = \int_{-\infty}^{x} \frac{1}{\sqrt{2\pi}}\,e^{-\frac{1}{2}t^2}\,dt \,,$$

being the area under $\phi(x)$ in the interval $(-\infty, x]$, see also Fig. 2.3[3]. The function $\Phi(x)$ cannot be expressed in terms of elementary functions, but the following is available.

Theorem 2.3 (Laplace, 1875) *The cumulative distribution function of the standard normal distribution $\Phi(x)$ can be written as*

$$\Phi(x) = \frac{1}{2} + \frac{1}{\sqrt{2\pi}} \sum_{n=0}^{\infty} (-1)^n \frac{x^{2n+1}}{n!(2n+1)2^n}$$

for $x \geq 0$.

The above power series expansion only holds for nonnegative x, but in view of the relation

$$\Phi(-x) = 1 - \Phi(x) \,,$$

it can also be applied to negative x. Nonetheless, it is known to converge rather slowly, and is therefore usually not used for computing $\Phi(x)$, see also [63, 3.3.1]. As an alternative, one may compute $\Phi(x)$ from the approximation due to Moran [56], given by

$$\Phi(x) \approx \frac{1}{2} + \frac{1}{\pi}\left[\frac{x}{3\sqrt{2}} + \sum_{n=1}^{12} \frac{1}{n}\,e^{-n^2/9}\sin\left(\frac{nx\sqrt{2}}{3}\right)\right], \quad -7 \leq x \leq 7\,,$$

This approximation is very easy to programme and precise enough for practical purposes, since it is accurate to 9 digits after the dot for any $-7 \leq x \leq 7$, see also [63, 3.3.3]. Note that there is hardly ever need

---

[3]Sometimes the function $\Phi(x)$ is introduced as the area under $\phi(x)$ in the interval $[0, x]$ for nonnegative x.

to compute $\Phi(x)$ for values x smaller than -3 or greater than $+3$, since then $\Phi(x)$ is almost identical to 0 and 1, respectively. Moreover, there is usually no need for the user to implement an algorithm on his own, since most statistical software packages provide a tool for computing $\Phi(x)$ for arbitrary x with sufficient accuracy. In addition, as a non-computer alternative, many statistical text books provide a table of values $\Phi(x)$ for $0 \le x \le 3$ and x having 4 digits after the dot, see Table 2.1.

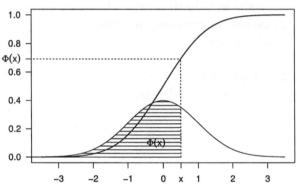

PDF and CDF of Standard Normal

Fig. 2.3: Cumulative distribution function $\Phi(x)$ (solid bold line) of the standard normal distribution together with corresponding probability density function $\phi(x)$

The importance of the function $\Phi(x)$ also lies in the fact that it can be used to compute the cumulative distribution function for any normal distribution.

Theorem 2.4 *Let X follow a $N(\mu, \sigma^2)$ distribution. Then the cumulative distribution function $F(x)$ of X is given by*

$$F(x) = \Phi[(x - \mu)/\sigma]$$

for $-\infty < x < \infty$.

x	0.00	0.01	0.02	0.03	0.04	0.05	0.06	0.07	0.08	0.09
0.0	0.5000	0.5040	0.5080	0.5120	0.5160	0.5199	0.5239	0.5279	0.5319	0.5359
0.1	0.5398	0.5438	0.5478	0.5517	0.5557	0.5596	0.5636	0.5675	0.5714	0.5753
0.2	0.5793	0.5832	0.5871	0.5910	0.5948	0.5987	0.6026	0.6064	0.6103	0.6141
0.3	0.6179	0.6217	0.6255	0.6293	0.6331	0.6368	0.6406	0.6443	0.6480	0.6517
0.4	0.6554	0.6591	0.6628	0.6664	0.6700	0.6736	0.6772	0.6808	0.6844	0.6879
0.5	0.6915	0.6950	0.6985	0.7019	0.7054	0.7088	0.7123	0.7157	0.7190	0.7224
0.6	0.7257	0.7291	0.7324	0.7357	0.7389	0.7422	0.7454	0.7486	0.7517	0.7549
0.7	0.7580	0.7611	0.7642	0.7673	0.7704	0.7734	0.7764	0.7794	0.7823	0.7852
0.8	0.7881	0.7910	0.7939	0.7967	0.7995	0.8023	0.8051	0.8078	0.8106	0.8133
0.9	0.8159	0.8186	0.8212	0.8238	0.8264	0.8289	0.8315	0.8340	0.8365	0.8389
1.0	0.8413	0.8438	0.8461	0.8485	0.8508	0.8531	0.8554	0.8577	0.8599	0.8621
1.1	0.8643	0.8665	0.8686	0.8708	0.8729	0.8749	0.8770	0.8790	0.8810	0.8830
1.2	0.8849	0.8869	0.8888	0.8907	0.8925	0.8944	0.8962	0.8980	0.8997	0.9015
1.3	0.9032	0.9049	0.9066	0.9082	0.9099	0.9115	0.9131	0.9147	0.9162	0.9177
1.4	0.9192	0.9207	0.9222	0.9236	0.9251	0.9265	0.9279	0.9292	0.9306	0.9319
1.5	0.9332	0.9345	0.9357	0.9370	0.9382	0.9394	0.9406	0.9418	0.9429	0.9441
1.6	0.9452	0.9463	0.9474	0.9484	0.9495	0.9505	0.9515	0.9525	0.9535	0.9545
1.7	0.9554	0.9564	0.9573	0.9582	0.9591	0.9599	0.9608	0.9616	0.9625	0.9633
1.8	0.9641	0.9649	0.9656	0.9664	0.9671	0.9678	0.9686	0.9693	0.9699	0.9706
1.9	0.9713	0.9719	0.9726	0.9732	0.9738	0.9744	0.9750	0.9756	0.9761	0.9767
2.0	0.9772	0.9778	0.9783	0.9788	0.9793	0.9798	0.9803	0.9808	0.9812	0.9817
2.1	0.9821	0.9826	0.9830	0.9834	0.9838	0.9842	0.9846	0.9850	0.9854	0.9857
2.2	0.9861	0.9864	0.9868	0.9871	0.9875	0.9878	0.9881	0.9884	0.9887	0.9890
2.3	0.9893	0.9896	0.9898	0.9901	0.9904	0.9906	0.9909	0.9911	0.9913	0.9916
2.4	0.9918	0.9920	0.9922	0.9925	0.9927	0.9929	0.9931	0.9932	0.9934	0.9936
2.5	0.9938	0.9940	0.9941	0.9943	0.9945	0.9946	0.9948	0.9949	0.9951	0.9952
2.6	0.9953	0.9955	0.9956	0.9957	0.9959	0.9960	0.9961	0.9962	0.9963	0.9964
2.7	0.9965	0.9966	0.9967	0.9968	0.9969	0.9970	0.9971	0.9972	0.9973	0.9974
2.8	0.9974	0.9975	0.9976	0.9977	0.9977	0.9978	0.9979	0.9979	0.9980	0.9981
2.9	0.9981	0.9982	0.9982	0.9983	0.9984	0.9984	0.9985	0.9985	0.9986	0.9986
3.0	0.9987	0.9987	0.9987	0.9988	0.9988	0.9989	0.9989	0.9989	0.9990	0.9990

Table 2.1: Values of $\Phi(x)$ for $0.00 \leq x \leq 3.09$, e.g. $\Phi(1.76) = 0.9608$

As already noted, the above theorem can be used for computing the cdf of any normally distributed random variable from $\Phi(x)$. Table 2.2 gives an overview about pdf and cdf of the normal distribution.

	pdf	cdf
$N(0,1)$	$\phi(x) = \frac{1}{\sqrt{2\pi}} e^{-\frac{1}{2}x^2}$	$\Phi(x) = \frac{1}{2} + \frac{1}{\sqrt{2\pi}} \sum_{n=0}^{\infty} (-1)^n \frac{x^{2n+1}}{n!(2n+1)2^n}, \ x \geq 0$
	$\phi(-x) = \phi(x)$	$\Phi(-x) = 1 - \Phi(x)$
	$\phi(0) = 1/\sqrt{2\pi}$	$\Phi(0) = 1/2$
$N(\mu,\sigma^2)$	$f(x) = \frac{1}{\sigma}\phi\left(\frac{x-\mu}{\sigma}\right)$	$F(x) = \Phi\left(\frac{x-\mu}{\sigma}\right)$

Table 2.2: Probability density and cumulative distribution function of the normal distribution

2.2.3   Related Functions

There exists a considerable amount of literature concerning the investigation of the functions $\phi(x)$, $\Phi(x)$ and related functions.

The Error Function

Related to the function $\Phi(x)$ is the so-called *error function*, given by

$$\mathrm{erf}(x) = \frac{2}{\sqrt{\pi}} \int_0^x e^{-t^2} \, dt \ .$$

The function $\Phi(x)$ can be expressed via the error function as

$$\Phi(x) = \frac{1}{2}(\mathrm{erf}(x/\sqrt{2}) + 1) \ .$$

Often, mathematical software packages (like MAPLE or Mathematica) provide an implementation of the error function, which can then be used to do computations concerning $\Phi(x)$.

Mills' Ratio

Another function related to $\phi(x)$ and $\Phi(x)$ is *Mills' ratio*, defined by

$$R(x) = \frac{1 - \Phi(x)}{\phi(x)} \ .$$

It occurs in the literature concerned with approximations to $\Phi(x)$, see [63, Chapter 3] for a discussion.

Normal Integrals

Eventually, it is interesting to note that there exist a number of formulas for so-called *normal integrals*, i.e. indefinite or definite integrals involving $\phi(x)$ and/or $\Phi(x)$, see [63, Sect. 2.5]. Examples for *indefinite normal integrals* are

$$\int \phi(x)\phi(a + bx)\, dx = \frac{1}{t}\phi\left(\frac{a}{t}\right)\Phi\left(tx + \frac{a}{t}\right), \quad t = \sqrt{1 + b^2} \ ,$$

$$\int x\phi(a + bx)\, dx = -\frac{1}{b^2}\phi(a + bx) - \frac{a}{b^2}\Phi(a + bx) \ ,$$

$$\int x^2\phi(a + bx)\, dx = \frac{a^2 + 1}{b^3}\Phi(a + bx) - \frac{bx - a}{b^3}\phi(a + bx) \ ,$$

$$\int \Phi(a + bx)\, dx = \frac{a + bx}{b}\Phi(a + bx) + \frac{1}{b}\phi(a + bx) \ ,$$

$$\int x\phi(x)\Phi(bx)\, dx = \frac{b}{\sqrt{2\pi(1 + b^2)}}\Phi(x\sqrt{1 + b^2}) - \phi(x)\Phi(bx) \ .$$

Examples for *definite normal integrals* are

$$\int_{-\infty}^{0} \phi(ax)\Phi(bx)\, dx = \frac{1}{2\pi a}\arctan\left(\frac{a}{b}\right) \ ,$$

$$\int_{0}^{\infty} \phi(ax)\Phi(bx)\, dx = \frac{1}{2\pi a}\left(\frac{\pi}{2} - \arctan\left(\frac{b}{a}\right)\right) \ ,$$

$$\int_{-\infty}^{\infty} x\phi(x)\Phi(bx)\, dx = \frac{b}{\sqrt{2\pi(1 + b^2)}} \ ,$$

$$\int_{-\infty}^{\infty} \Phi(a + bx)\phi(x)\, dx = \Phi\left(\frac{a}{\sqrt{1 + b^2}}\right) \ .$$

Setting $a = 0$ in the last formula gives $\int_{-\infty}^{\infty} \Phi(bx)\phi(x)\, dx = \Phi(0) = 1/2$, showing that the function $2\Phi(bx)\phi(x)$ is a pdf for any choice of b. As a matter of fact, this is the pdf of the so-called *skew-normal* distribution.

Normal integrals also turn out to be useful for the computation of expectation and variance of random variables having a distribution related to the normal, and some normal integrals are applied in this text. See also [58, 13] for an extensive list of normal integrals.

2.3   Moments

As noted before, for a $N(\mu, \sigma^2)$ distributed random variable, expectation and variance are $E(X) = \mu$ and $\text{Var}(X) = \sigma^2$, respectively.

2.3.1   The 1, 2, 3 standard deviation intervals

From the Chebyshev inequality we can obtain a lower bound for any random variable X to realize in the interval

$$(\mu - k\sigma, \mu + k\sigma)$$

where $\mu = E(X)$, $\sigma = \sqrt{\text{Var}(X)}$ and $k > 0$. As already noted in Sect. 1.2.2, this lower bound is often too small, i.e. the actual probability for X to realize in this interval will be greater than this bound. When X is $N(\mu, \sigma^2)$ distributed, then we have

$$P[X \in (\mu - k\sigma, \mu + k\sigma)] = F(\mu + k\sigma) - F(\mu - k\sigma) \,,$$

where $F(x)$ is the cdf of $N(\mu, \sigma^2)$. By applying Theorem 2.4, it follows that

$$P[X \in (\mu - k\sigma, \mu + k\sigma)] = \Phi(k) - \Phi(-k) = 2\Phi(k) - 1 \,.$$

For the choices $k = 1, 2, 3$ we have

$$P[X \in (\mu - \sigma, \mu + \sigma)] = 0.6827 \,,$$

$$P[X \in (\mu - 2\sigma, \mu + 2\sigma)] = 0.9545 \,,$$

and

$$P[X \in (\mu - 3\sigma, \mu + 3\sigma)] = 0.9973 \,.$$

This shows that more than 95% of the outcomes of a normally distributed random variable will be in the 2σ interval around μ, and about 99.73% will be in the 3σ interval around μ.

Real-life Variables and Normality

The high probability of the 3σ interval can also be seen as a justification for the possible assumption of normality of real-life variables. In general, a real-life variable, like the weight of person, can only take values in a certain interval (e.g. the weight of a person cannot take negative values). But this is not true for a normally distributed random variable (e.g. the probability for a normally distributed random variable to come out in the interval $(-\infty, 0)$ is always positive, irrespective of μ and σ). Hence, strictly speaking, such a real-life variable cannot be normally distributed. On the other hand, if μ and σ of a real-life variable X can be assumed to satisfy the requirement such that the 3σ interval around μ does not exceed the natural bounds of this variable, then it is nonetheless acceptable to assume the normality of this variable, since under normality almost all (99.73%) outcomes lie in the 3σ interval[4].

Nonetheless, it is better style to say that 'the weight of a person *is assumed* to be normally distributed' than to say 'the weight of a person *is* normally distributed'.

2.3.2 Higher Moments

So far we have mainly considered expectation and variance as two characteristic numbers of a random variable X. These turn out to be special non-central and central moments, respectively.

Definition 2.2 *Let X be a random variable. Then the k-th non-central moment of X is*

$$\mathrm{E}(X^k)$$

and the k-th central moment of X is

$$\mu_k := \mathrm{E}[(X - \mathrm{E}(X))^k]$$

for $k = 1, 2, 3, \ldots$.

---

[4]Of course, this does not mean that any variable producing more than 99% of observations in the 3σ interval around μ has necessarily a normal distribution. The normality assumption should be based on good reasons. See also Chapters 3 and 4 for possible ways to check for normality.

For any random variable X, its expectation is its first non-central moment $E(X)$, and its variance is its second central moment μ_2. Since the $N(\mu, \sigma^2)$ distribution is entirely determined by μ and σ, higher moments naturally come out as functions of these two parameters.

Normal Non-Central Moments

When X is $N(\mu, \sigma^2)$ distributed, then for the non-central moments $E(X^k)$ we have

$$E(X^{2r-1}) = \sigma^{2r-1} \sum_{i=1}^{r} \frac{(2r-1)!}{(2i-1)!(r-i)!2^{r-i}} \left(\frac{\mu}{\sigma}\right)^{2i-1}, \quad r = 1, 2, \ldots,$$

and

$$E(X^{2r}) = \sigma^{2r} \sum_{i=0}^{r} \frac{(2r)!}{(2i)!(r-i)!2^{r-i}} \left(\frac{\mu}{\sigma}\right)^{2i}, \quad r = 1, 2, \ldots.$$

To compute $E(X^{r+1})$ from $E(X^r)$ and $E(X^{r-1})$, one may also apply the formula

$$E(X^{r+1}) = r\sigma^2 E(X^{r-1}) + \mu E(X^r), \quad r = 1, 2, 3, \ldots.$$

Normal Central Moments

When X is $N(\mu, \sigma^2)$ distributed, then for the central moments $\mu_k = E[(X - E(X))^k]$ we have

$$\mu_{2r+1} = 0, \quad r = 0, 1, 2, \ldots,$$

and

$$\mu_{2r} = \frac{(\sigma^2/2)r(2r)!}{r!}, \quad r = 0, 1, 2, \ldots.$$

The first four central moments of $N(\mu, \sigma^2)$ are $\mu_1 = \mu$, $\mu_2 = \sigma^2$, $\mu_3 = 0$, and $\mu_4 = 3\sigma^4$.

2.3.3   Skewness and Kurtosis

When the pdf of a distribution is symmetric, the odd central moments can be shown to be equal to zero. Hence, also $\mu_3 = 0$, being often called

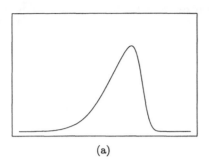 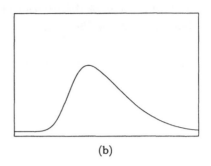

(a)   (b)

Fig. 2.4: Probability density function skewed to the left (a) and skewed to the right (b)

a measure of skewness. If a pdf is skewed to the left as in Fig. 2.4 (a), then $\mu_3 < 0$, while for a pdf skewed to the right as in Fig. 2.4 (b) we have $\mu_3 > 0$.

However, there exist non-symmetric distributions with $\mu_3 = 0$, see e.g. Exercise 3.26 in [83], so that actually the value of μ_3 does not necessarily give a hint to the shape of the pdf. Nonetheless, in the literature,

$$\sqrt{\beta_1} := \mu_3/\mu_2^{3/2} = \mu_3/\sigma^3 \,,$$

where $\mu_2 = \sigma^2$, is usually referred to as the *coefficient of skewness*. For $N(\mu, \sigma^2)$ we have $\sqrt{\beta_1} = 0$.

A characteristic number involving the fourth central moment of a random variable is the *kurtosis*, given by

$$\beta_2 := \mu_4/\mu_2^2 = \mu_4/\sigma^4 \,.$$

The $N(\mu, \sigma^2)$ distribution has kurtosis $\beta_2 = 3$. The so-called *coefficient of excess* $\beta_2 - 3$ compares the kurtosis of a given pdf with the kurtosis of the normal pdf. The kurtosis (or the coefficient of excess) is often applied for comparison purposes when the given pdf is symmetric. Then values of the kurtosis greater than 3 are used to indicate a more peaked pdf than the normal, also often having heavier tails. As an example, consider the pdf

$$f(x) = \frac{\pi}{\sqrt{3}} \frac{e^{-\pi x/\sqrt{3}}}{(1 + e^{-\pi x/\sqrt{3}})^2}, \quad -\infty < x < \infty \,,$$

(standardized logistic distribution) having expectation 0, variance 1, $\sqrt{\beta_1} = 0$, but $\beta_2 = 4.2$, see Fig. 2.5.

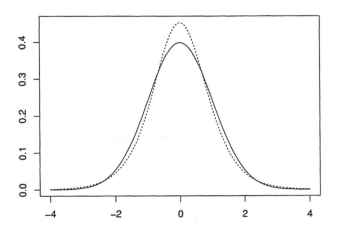

Fig. 2.5: Probability density function of standard normal (solid line) compared with the pdf of the logistic distribution (dotted line), having expectation 0 and variance 1

In general, however, the meaning of kurtosis is not all clear. There exist different pdfs with the same kurtosis and there also exist non-normal pdfs with a kurtosis of 3, see also [7].

Remark 2.2 In view of $\mu_k(a + bX) = b^k \mu_k(X)$ for scalars a and b, it follows for any random variable X that

$$\sqrt{\beta_1}(a + bX) = \sqrt{\beta_1}(X) \quad \text{and} \quad \beta_2(a + bX) = \beta_2(X) \,,$$

i.e. $\sqrt{\beta_1}$ and β_2 are invariant under linear transformations of a random variable.

2.3.4   Sample Skewness and Kurtosis

For a sample $X_1, \ldots, X_n$ consider the *j-th central sample moment*

$$m_j = \frac{1}{n} \sum_{i=1}^{n} (X_i - \overline{X})^j, \quad j = 2, 3, 4, \ldots .$$

This definition of a sample moment is in a slight contradiction with the definition of the sample variance, since one would expect the second sample moment m_2 to be identical to the sample variance S^2, while it is actually identical to D^2. It is of course possible to define D^2 as the sample variance (and it is sometimes done), but the notions of sample variance and sample moment used here are more common. Now, the *sample skewness* is

$$\sqrt{b_1} = m_3/m_2^{3/2}$$

and the *sample kurtosis* is

$$b_2 = m_4/m_2^2 .$$

Sample skewness and kurtosis are random variables whose exact distribution is still not known, even under a sample from the normal distribution. However, we may state the following.

Theorem 2.5 *Let $X_1, \ldots, X_n$ be a sample from the $N(\mu, \sigma^2)$ distribution. Then*

$$\mathrm{E}(\sqrt{b_1}) = 0 \quad and \quad \mathrm{Var}(\sqrt{b_1}) = 6\frac{n-2}{(n+1)(n+3)},$$

aa well as

$$\mathrm{E}(b_2) = 3\frac{n-1}{n+1} \quad and \quad \mathrm{Var}(b_2) = 24\frac{n(n-2)(n-3)}{(n+1)^2(n+3)(n+5)} .$$

Moreover, $\sqrt{b_1}$ and b_2 are uncorrelated but not independent.

Considerable deviations of the sample skewness from 0 and/or the sample kurtosis from 3 can indicate that the given sample does not come from a normal distribution.

2.4 Quantiles

The cumulative distribution function $F(x)$ of a random variable X gives the probability for X to take a value in the interval $(-\infty, x]$, meaning that

$$F(x) = \mathrm{P}(X \le x) \ .$$

Suppose we are interested in the converse question: given a value $p \in (0,1)$, for what x does $\mathrm{P}(X \le x) = p$ hold? Such an x associated with a given probability p is called a p-*quantile* of X and is denoted by ξ_p, so that

$$\mathrm{P}(X \le \xi_p) = p \ .$$

In general such a ξ_p must not exist, since the cdf $F(x) \in [0,1]$ of an arbitrary random variable does not necessarily take every value in $[0,1]$. This is for example the case when X is discrete, in which case $F(x)$ is a step function, taking only each specific step in $[0,1]$ as a value.

When X is a continuous random variable, existence of such a ξ_p is guaranteed, since the cdf of X is continuous. But it may not be unique[5], i.e. there may exist more than one ξ_p satisfying the relationship $\mathrm{P}(X \le \xi_p) = p$ for a given p. This is due to the fact that the cdf of a continuous random variable is not necessarily *strictly* monotone. As an example consider the pdf

$$f(x) = \frac{1}{3(b-a)} I_{[a,b]}(x) + \frac{2}{3(d-c)} I_{[c,d]}(x), \quad a < b < c < d \ ,$$

where a p-quantile for $p = 1/3$ is any number between b and c, compare Fig. 1.7 (a) and (b), p. 14.

2.4.1   Quantiles of the Normal Distribution

For the $N(\mu, \sigma^2)$ distribution, however, the cdf is strictly monotone and thus there exists a unique ξ_p satisfying $\mathrm{P}(X \le \xi_p) = p$ for every $p \in (0,1)$. The p-quantile divides the area beneath the pdf of X such that it equals p in $(-\infty, \xi_p]$ and $1 - p$ in $[\xi_p, \infty)$, see Fig. 2.6. In view of the identity

$$\Phi(x) = F(\mu + \sigma x) \ ,$$

where $F(\cdot)$ is the cdf of the $N(\mu, \sigma^2)$ distribution, the p-quantiles of the $N(\mu, \sigma^2)$ distribution can be computed from the corresponding p-quantiles of the $N(0,1)$ distribution.

---

[5]To guarantee existence and uniqueness, one may define the p-quantile of a random variable X to be the smallest number ξ satisfying $\mathrm{P}(X \le \xi) \ge p$.

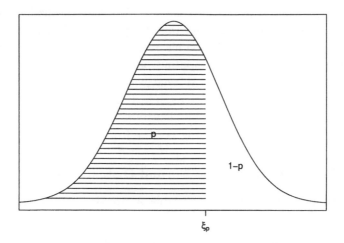

Fig. 2.6: p-quantile of the normal pdf

Theorem 2.6 *Let u_p denote the p-quantile of the standard normal distribution. Then the p-quantile of the $N(\mu, \sigma^2)$ distribution is given by $\xi_p = \mu + \sigma u_p$.*

The p-quantile u_p of the standard normal distribution is the unique solution to the equation $p = \Phi(u)$ with respect to u. We may also write

$$u_p = \Phi^{-1}(p) \,,$$

where $\Phi^{-1} : (0, 1) \to \mathbb{R}$ may be called the *quantile function* of the standard normal distribution. For the actual computation of a value $u_p = \Phi^{-1}(p)$ for some given $p \in (0, 1)$ we have the same problem as with the cdf $\Phi(\cdot)$, namely the quantile function cannot be expressed via elementary functions. In view of the relation

$$u_p = -u_{1-p} \,,$$

it is enough to know u_p for $0.5 \leq p < 1$. Now, for the actual computation one may use the approximation due to Hastings [31, pp. 191/912], given as

$$u_p \approx t - \frac{2.515517 + 0.802853t + 0.010328t^2}{1 + 1.432788t + 0.189269t^2 + 0.001308t^3}, \quad 0.5 \leq p < 1 \,,$$

where $t = \sqrt{-2\ln(1-p)}$. See also [63, Sect. 3.9] for a discussion of further approximation formulas, but note that there is usually no need for the user to implement an algorithm on his own, since most statistical software packages provide a tool for computing p-quantiles of the normal distribution with sufficient accuracy. Alternatively, many statistical text books provide a table of values u_p for $0.5 \le p < 1$. One may also obtain such a quantile from Table 2.1, p. 53, by identifying a value p within the table and then deduce the corresponding x-value from the margin of the table.

Median and Quartiles of the Normal Distribution

Of specific interest of a random variable X are the p-quantiles ξ_p for the choices

- $p = 0.5$, the p-quantile being called the *median* of X,
- $p = 0.25$ the p-quantile being called the *lower quartile* of X, and
- $p = 0.75$, the p-quantile being called the *upper quartile* X.

For the $N(0,1)$ distribution we have $u_{0.5} = 0$, $u_{0.75} = 0.6744898$, and $u_{0.25} = -u_{0.75}$.

Corollary 2.1 *Let X be $N(\mu, \sigma^2)$ distributed. Then*

$$\xi_{0.5}(X) = \mu, \quad \xi_{0.25}(X) = \mu + \sigma u_{0.25}, \quad and \quad \xi_{0.75}(X) = \mu + \sigma u_{0.75} .$$

When $\xi_{0.5}$ is the median of a random variable X, let MAD (median of the absolut deviation from the median) of X be the median of $|X - \xi_{0.5}|$. Then for X being $N(\mu, \sigma^2)$ distributed and written as $X = \mu + \sigma Z$, where Z has standard normal distribution, it can be shown that

$$\mathrm{MAD}(X) = \xi_{0.5}(\sigma|Z|) = \sigma u_{0.75} \approx \sigma 0.6745 .$$

When we consider the interquartile range IQR $= \xi_{0.75} - \xi_{0.25}$ of the $N(\mu, \sigma^2)$ distribution, then it follows

$$\mathrm{IQR} = (u_{0.75} - u_{0.25})\sigma = 2u_{0.75}\sigma \approx 1.34\sigma .$$

Hence, for X being $N(\mu, \sigma^2)$ distributed it follows that $\mathrm{MAD}(X) = \mathrm{IQR}(X)/2$.

2.4.2 Sample Quantiles

Corresponding to a quantile of a given pdf, there is a quantile of a given sample, defined with respect to the *ordered* sample.

Order Statistics

If $x_1, \ldots, x_n$ are the observations of a sample, then each observation x_i is regarded as the outcome of an individual random variable X_i, where $X_1, \ldots, X_n$ are independent and identically distributed. As already noted, a *statistic* is defined to be a function of the sample $X_1, \ldots, X_n$, so that the minimum $X_{(1)} = \min_{1 \leq i \leq n} X_i$ is a statistic, also called the *first order statistic*. When we consider the minimum $X_{(2)}$ of the remaining X_i (without $X_{(1)}$), then this is also called the *second order statistic*. Proceeding in this way, $X_{(i)}$ is called the *i-th order statistic* of the sample $X_1, \ldots, X_n$, the order statistics satisfying

$$X_{(1)} \leq X_{(2)} \leq \cdots \leq X_{(n)} .$$

When we order our observations $x_1, \ldots, x_n$ such that $x_{(1)} \leq x_{(2)} \leq \cdots \leq x_{(n)}$, then the i-th order statistic $X_{(i)}$ can be seen as the random variable standing behind $x_{(i)}$. As opposed to the X_i, the $X_{(i)}$ are usually not independent.

If the sample $X_1, \ldots, X_n$ comes from a continuous distribution with pdf $f(x)$ and cdf $F(x)$, then the pdf of $X_{(i)}$ is given as

$$f_{(i)}(x) = \frac{n!}{(i-1)!\,(n-i)!} f(x)[F(x)]^{i-1}[1 - F(x)]^{n-i} ,$$

see e.g. [12, Theorem 5.4.4]. Hence, if the sample comes from the normal distribution with parameters μ and σ^2, the pdf of the i-th *normal order statistic* is $f_{(i)}(x)$ with

$$f(x) = \frac{1}{\sigma}\phi\left(\frac{x-\mu}{\sigma}\right) \quad \text{and} \quad F(x) = \Phi\left(\frac{x-\mu}{\sigma}\right) .$$

To obtain moments $\mathrm{E}(X_{(i)})$ and $\mathrm{E}(X_{(i)}^2)$, it is enough to consider the moments of the order statistics $Z_{(1)}, \ldots, Z_{(n)}$, obtained from a standard normal sample $Z_1, \ldots, Z_n$, since

$$\mathrm{E}(X_{(i)}) = \mu + \sigma\mathrm{E}(Z_{(i)}) \quad \text{and} \quad \mathrm{E}(X_{(i)}^2) = \mu^2 + 2\mu\sigma\mathrm{E}(Z_{(i)}) + \sigma^2\mathrm{E}(Z_{(i)}^2) .$$

For arbitrary sample sizes n, the moments $\mathrm{E}(Z_{(i)})$ and $\mathrm{E}(Z_{(i)}^2)$ cannot be given explicitly[6], but may be computed from a numerical evaluation of the integrals $\int_{-\infty}^{\infty} x f_{(i)}(x)\,\mathrm{d}x$ and $\int_{-\infty}^{\infty} x^2 f_{(i)}(x)\,\mathrm{d}x$, respectively, where $\mu = 0$ and $\sigma = 1$ in $f_{(i)}(x)$, see e.g. [61, 62]. A well known approximation formula from Blom [8] is

$$\mathrm{E}(Z_{(i)}) = \Phi^{-1}\left(\frac{i - \delta}{n - 2\delta + 1}\right)$$

for some $0 \le \delta < 1$. Blom suggested the choice $\delta = 3/8$, while Harter [30] discussed more appropriate choices for δ, depending on the sample size n. Nonetheless, with the aid of computer based numerical integration procedures, it is possible for the user to evaluate the above mentioned integrals on his own.

Sample Quantiles

Let $X_1, \ldots, X_n$ be a sample, and let $X_{(1)}, \ldots, X_{(n)}$ be the corresponding ordered sample. Then the *sample median* is defined as

$$M = \begin{cases} X_{((n+1)/2)} & \text{if } n \text{ is odd} \\ \frac{1}{2}(X_{(n/2)} + X_{(n/2+1)}) & \text{if } n \text{ is even} \end{cases}.$$

The sample median is a random variable, being a function of order statistics. The actual value of the sample median is obtained from the ordered observations when upper case $X_{(i)}$'s are replaced by observed lower case $x_{(i)}$'s.

Since the median is known to be a special quantile, one may also define sample quantiles. If np is not an integer, then the *sample p-quantile* $\widehat{\xi}_p$ is defined as

$$\widehat{\xi}_p = X_{([np]+1)} ,$$

where $[np]$ is the integer part of np. If np is an integer, then the sample p-quantile can be defined to be any number

$$\widehat{\xi}_p \in [X_{(np)}, X_{(np+1)}] .$$

---

[6]For sample sizes $n = 2, \ldots, 5$, the exact moments $\mathrm{E}(Z_{(i)})$ and $\mathrm{E}(Z_{(i)}^2)$ are available and given e.g. in [63, Table 8.2]

As can be seen from the definition of the sample median, if n is even, the sample 0.5-quantile is usually chosen as the arithmetic mean between $X_{(np)}$ and $X_{(np+1)}$. For the actual computation of the p-quantile from given values $x_1, \ldots, x_n$ by a statistical software package, one may find slight differences depending on the used product. These differences, however, are usually only marginal and not of practical relevance.

The sample version of MAD is computed as the sample median of the transformed sample $Y_1, \ldots, Y_n$, $Y_i = |X_i - M|$, where M is the sample median of the original sample $X_1, \ldots, X_n$. The sample version of IQR is computed as the difference $\widehat{\xi}_{0.75} - \widehat{\xi}_{0.25}$.

2.5 Parameter Estimation

When we assume data points $x_1, \ldots, x_n$ to come from a normal distribution, then usually both parameters μ and σ^2 are not known. Hence, for fitting $N(\mu, \sigma^2)$ to the data, the parameters must both be estimated.

2.5.1 Maximum Likelihood Estimators

Let $X_1, \ldots, X_n$ be a sample from the $N(\mu, \sigma^2)$ distribution. For the purpose of deriving an estimator for the two-dimensional parameter $\theta = (\mu, \sigma^2)$, we consider the likelihood function, being the joint pdf of the random variables $X_1, \ldots, X_n$ regarded as a function of θ. Hence,

$$L(\theta; x_1, \ldots, x_n) = \prod_{i=1}^{n} f(x_i; \theta) = \frac{1}{(\sigma^2 2\pi)^{n/2}} e^{-\frac{1}{2} \sum_{i=1}^{n} (x_i - \mu)^2 / \sigma^2} .$$

The log-likelihood is given as

$$\ln [L(\theta; x_1, \ldots, x_n)] = -\frac{n}{2} \ln(2\pi) - \frac{n}{2} \ln(\sigma^2) - \frac{1}{2} \sum_{i=1}^{n} (x_i - \mu)^2 / \sigma^2 .$$

Considering the partial derivatives

$$\frac{\partial}{\partial \mu} \ln [L(\theta; x_1, \ldots, x_n)] = \frac{1}{\sigma^2} \sum_{i=1}^{n} (x_i - \mu)$$

and

$$\frac{\partial}{\partial\sigma^2}\ln\left[L(\theta;x_1,\ldots,x_n)\right] = -\frac{n}{2\sigma^2} + \frac{1}{2\sigma^4}\sum_{i=1}^{n}(x_i-\mu)^2\,,$$

and setting both equal to zero, yields

$$\widehat{\mu} = \frac{1}{n}\sum_{i=1}^{n}x_i = \overline{x} \quad \text{and} \quad \widehat{\sigma}^2 = \frac{1}{n}\sum_{i=1}^{n}(x_i-\overline{x})^2 = d^2$$

as solutions with respect to μ and σ^2. Hence $\widehat{\theta} = (\widehat{\mu},\widehat{\sigma}^2) = (\overline{x},d^2)$ is a possible candidate for giving the global maximum of the likelihood function with respect to θ. As a matter of fact, $\widehat{\theta}$ *is* the global maximizer, see Examples 7.2.11 and 7.2.12 in [12] for a discussion on how to verify this. This shows that the corresponding maximum likelihood estimator for θ is $(\overline{X}, D^2)$. Then, in view of Theorem 1.4 we may state the following.

Theorem 2.7 *Let $X_1,\ldots,X_n$ be a sample from the $N(\mu,\sigma^2)$ distribution with unknown parameters μ and σ^2. Then $\overline{X}$, D^2, and $D = \sqrt{D^2}$ are the maximum likelihood estimators of μ, σ^2, and σ, respectively.*

2.5.2   Properties of Estimators

As noted before, the maximum likelihood approach is most common for obtaining estimators. In the following we shortly discuss properties of estimators for μ and σ^2 for a sample from the $N(\mu,\sigma^2)$ distribution, giving also some comments concerning the sample size n and the possible application of the estimators for non-normal samples.

Estimation of μ

The maximum likelihood estimator $\overline{X}$ for μ is unbiased for μ having variance σ^2/n. It is known, cf. [44, Chapt. 2, Example 2.1], that there is no other unbiased estimator having smaller variance, so that $\overline{X}$ is in fact the UMVUE for μ.

As an example of an alternative estimator, one might consider the sample median M, being also unbiased for μ. However, as noted from Theorem 2.11, the variance of M for larger sample sizes n is $(\pi/2)(\sigma^2/n)$, and thus greater than the variance of $\overline{X}$. This is of course in accordance with the fact that $\overline{X}$ is the UMVUE for μ.

Nonetheless, the sample median can be a serious competitor to the sample mean when the sample $X_1, \ldots, X_n$ does not come from the normal distribution, but for example from a distribution producing outliers, and it is desired to estimate the expectation μ of this distribution. In such a case the sample median is more robust and thus may give better estimates. For non-normal samples, the sample mean $\overline{X}$ remains unbiased for the expectation μ having variance σ^2/n, see Theorem 1.2, but it is not necessarily UMVUE for μ, so that possibly better estimators exist. On the other hand, for larger sample sizes, reasonable competitors to $\overline{X}$ should be consistent estimators with the same or better convergence rate towards μ.

Estimation of σ^2

The maximum likelihood estimator D^2 is not unbiased for σ^2. Hence, often the alternative $S^2 = \frac{n}{n-1}D^2$ is used, being in fact the UMVUE for σ^2, cf. [44, Chapt. 2, Example 2.1]. As a further alternative one may consider the estimator

$$D_*^2 = \frac{1}{n+1}\sum_{i=1}^{n}(X_i - \overline{X})^2 \,,$$

being identical to λS^2 with $\lambda = \frac{n+1}{n-1}$. It can be shown that this choice of λ yields the minimal mean squared error estimator among all estimators of the form λS^2, i.e.

$$\min_{\lambda} \mathrm{E}[(\lambda S^2 - \sigma^2)^2] = \mathrm{E}[(D_*^2 - \sigma^2)^2] \,.$$

Nonetheless it can be argued that the mean squared error criterium is not appropriate for assessing the performance of an estimator for σ^2, since it does not take into account the non-symmetric parameter space $(0, \infty)$. Under alternative criteria it can even be shown that S^2 itself performs best, see Examples 7.3.26 and 7.3.27 in [12].

For a larger sample size n, the difference between the three estimators S^2, D^2, and D_*^2 is only marginal, so that the actual choice of estimator has almost no practical relevance, each one producing nearly the same estimate.

For non-normal samples, often S^2 is preferred in view of its unbiasedness and consistency properties.

Estimation of σ

When estimation of the standard estimation σ is of interest, the sample standard estimator $S = \sqrt{S^2}$ is not unbiased for σ. As a matter of fact it can be shown that

$$E(S) = c_n\sigma, \quad c_n = \frac{\Gamma(\frac{n}{2})}{\sqrt{\frac{n-1}{2}}\Gamma(\frac{n-1}{2})},$$

where $\Gamma(\cdot)$ is the *gamma function* defined by

$$\Gamma(x) = \int_0^\infty t^{x-1}e^{-t}\,dt, \quad x > 0.$$

Hence, an unbiased estimator for σ is given by

$$S_u = c_n^{-1}S = \frac{\Gamma(\frac{n-1}{2})}{\sqrt{2}\Gamma(\frac{n}{2})}\sqrt{\sum_{i=1}^n (X_i - \overline{X})^2}.$$

As a matter of fact, this estimator is known to be the UMVUE for σ, cf. [44, Chap. 2, Example 2.1]. Nonetheless, the estimator S_u is rarely used, the estimator S, though biased, being quite more common. One reason for this is that even for smaller n the quantity c_n is not far away from 1, e.g. $c_{10} = 0.9727$, $c_{20} = 0.9869$, $c_{30} = 0.9914$, and $c_\infty = 1$.

The estimator S is also the most common estimator for the standard deviation from a non-normal sample. As an alternative, sometimes the sample mean absolute deviation from the median

$$\frac{1}{n}\sum_{i=1}^n |X_i - M|$$

is applied, being more robust with respect to outliers. Other robust estimators for σ are derived from normal properties. For the $N(\mu, \sigma^2)$ distribution we have

$$\sigma \approx \text{MAD}/0.6745 \quad \text{as well as} \quad \sigma \approx \text{IQR}/1.34.$$

By considering the respective sample versions, the right-hand sides are also used as robust estimators for the standard deviation for non-normal samples.

2.6 The Central Limit Theorem

Essentially, the central limit theorem claims that for large sample size n, the distribution of the sample mean $\overline{X}_n$ can always be approximated by a specific normal distribution, irrespective of the distribution of the sample $X_1, \ldots, X_n$. Patel and Read [63, Chapter 6] discuss a variety of different forms of central limit theorems.

2.6.1 Approximate Normality of the Sample Mean

We cite the *central limit theorem* (CLT) in its form as given in [12, Theorem 5.5.15].

Theorem 2.8 *Let $X_1, \ldots, X_n, \ldots$ be a sequence of independent and identically distributed random variables with $\mathrm{E}(X_i) = \mu$ and $0 < \mathrm{Var}(X_i) = \sigma^2 < \infty$. Let $\overline{X}_n = \frac{1}{n} \sum_{i=1}^n X_i$ and let $G_n(x)$ be the cumulative distribution function of the random variable $\sqrt{n}(\overline{X}_n - \mu)/\sigma$. Then for any x, $-\infty < x < \infty$,*

$$\lim_{n \to \infty} G_n(x) = \int_{-\infty}^x \frac{1}{\sqrt{2\pi}} e^{-\frac{1}{2}t^2} \, dt \, ,$$

that is $\sqrt{n}(\overline{X}_n - \mu)/\sigma$ has a limiting standard normal distribution.

The interesting fact about the central limit theorem is, that it requires no assumptions about the actual distribution of the $X_1, \ldots, X_n, \ldots$, other than that of finite variance $\sigma^2 < \infty$.

The practical relevance lies in the fact that whenever we are confronted with a mean $\frac{1}{n} \sum_{i=1}^n X_i$ or a sum $\sum_{i=1}^n X_i$ of independent and identically distributed random variables $X_1, \ldots, X_n$ with arbitrary distribution, then we may say that it has an *approximate normal distribution*. (We use the term 'approximate' when we talk about a fixed sample size n.)

The mean $\frac{1}{n} \sum_{i=1}^n X_i$ has an approximate normal distribution with expectation μ and variance σ^2/n, while the sum has an approximate normal distribution with mean $n\mu$ and variance $n\sigma^2$. The order of magnitude of finite n, guaranteeing a sufficient degree of approximation, however, does in fact depend on the actual distribution of the X_i. In general, when the distribution of the X_i is symmetric, the approximation can work well even

for small n, while for an asymmetric distribution of the X_i, the approximation will require greater values of n, the more pronounced the asymmetry
the greater the required n.

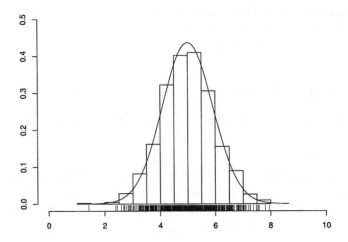

Fig. 2.7: Histogram of 1000 values of $Y = \sum_{i=1}^{n} X_i$ and pdf of the
$N(n/2, n/12)$ distribution for $n = 10$

Example 2.1 Consider a sample $X_1, \ldots, X_n$ from the pdf

$$f(x) = \begin{cases} 1/(b-a) & \text{if } a \leq x \leq b \\ 0 & \text{otherwise} \end{cases}$$

for $a = 0$ and $b = 1$ (uniform distribution on $[a, b]$). Then $\mu = \mathrm{E}(X_i) = (a+b)/2 = 1/2$ and $\sigma^2 = \mathrm{Var}(X_i) = (b-a)^2/12 = 1/12$. When we consider
the distribution of $Y := \sum_{i=1}^{n} X_i$, then this cannot be normal, since there
is a zero probability for Y to realize in $(-\infty, 0)$ and (n, ∞), while there is a
strictly positive probability for a normally distributed random variable to
realize within these intervals[7]. Nonetheless, even for smaller values of n, the
$N(n/2, n/12)$ distribution appears to be a good approximation to the true

---

[7]Although this probability can be quite small, depending on the location and scale
of the normal distribution.

distribution of Y. To illustrate this, we generate 1000 samples of size $n = 10$ from the above uniform distribution and compute the respective values of Y. Fig. 2.7 shows a histogram of these values together with the density of the $N(n/2, n/12)$ distribution for $n = 10$. Note that the 3σ interval for the $N(n/2, n/12)$ distribution is given by $(n/2 - 3\sqrt{n/12}, n/2 + 3\sqrt{n/12})$, i.e. $(2.2614, 7.7386)$ for $n = 10$. For our 1000 values of Y we have 997 within this interval, being exactly the number we can expect under normality. $\square$

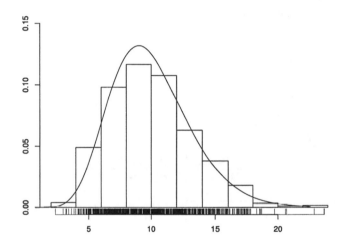

Fig. 2.8: Histogram of 1000 values of $Y = \sum_{i=1}^{n} X_i$ for $n = 10$

Example 2.2 Consider a sample $X_1, \ldots, X_n$ from the pdf

$$f(x) = \begin{cases} \lambda e^{-\lambda x} & \text{if } 0 < x < \infty \\ 0 & \text{otherwise} \end{cases}$$

(exponential distribution) for the choice $\lambda = 1$. Then $\mu = E(X_i) = 1/\lambda = 1$ and $\sigma^2 = \text{Var}(X_i) = 1/\lambda^2 = 1$. When we consider the distribution of $Y := \sum_{i=1}^{n} X_i$, then again this cannot be normal, since there is a zero probability for Y to realize in $(-\infty, 0)$. Moreover, we even know the correct distribution of Y, being the gamma distribution with pdf

$$f(x) = \frac{\lambda^n}{\Gamma(n)} x^{n-1} e^{-\lambda x} I_{(0,\infty)}(x) \ .$$

Figure 2.8 shows a histogram of 1000 values of Y, each obtained from a sample of size $n = 10$ from the above exponential distribution, together with the pdf of the gamma distribution. As can be seen, the $N(n, n)$ distribution would not provide a good approximation, since the distribution of Y is visibly skewed. Nonetheless for greater n, the gamma distribution with parameters n and λ approaches the $N(n/\lambda, n/\lambda^2)$ distribution, and then the approximation becomes better, see Fig. 2.9. □

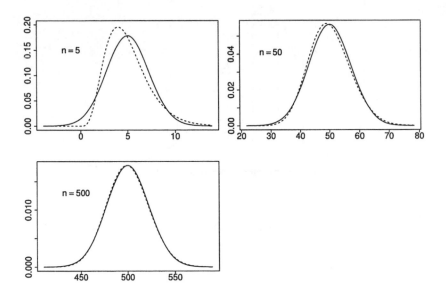

Fig. 2.9: Probability density functions of the $N(n/\lambda, n/\lambda^2))$ distribution (solid line) and the gamma distribution with parameters n and λ (dotted line) for the cases $n = 5$, $n = 50$ and $n = 500$ with $\lambda = 1$

There are a number of applications for the central limit theorem. For example, test statistics are often based on a sum or mean of the sample. Then for sufficiently large n the distribution of the test statistic can be approximated by a normal distribution. For different tests, there exist rules of thumb for the order of magnitude of n.

Often, when a random variable represents a (usually non-observable) random error, it is assumed that this error is the aggregation of a large number of different small errors, none of which dominating the others. Then the central limit theorem is taken as a justification for the assumption of normality of the random error. Although this may work well in many cases, there exist nonetheless instances when an error variable has its own *non-normal* distribution.

2.6.2 Cautionary Notes

As becomes evident from Examples 2.1 and 2.2, the central limit theorem itself does not claim about the goodness of approximation to normality with respect to the sample size n. Hence, in practice it may very well happen that the sample size is too small to admit a good approximation.

In addition, it is well known that the central limit theorem does not apply in any case. Although its assumptions may be weakened in one or the other direction, they cannot completely be neglected. The most famous example in this connection is a sample $X_1, \ldots, X_n$, coming from a distribution with pdf

$$f(x) = \frac{1}{\pi b} \frac{1}{(x-a)^2/b^2 + 1}$$

for some a and $b > 0$. This is the pdf of the *Cauchy distribution*[8]. It is known, that the sample mean $\overline{X}$ has this same Cauchy distribution irrespective of the sample size n, see [54, Sect. VI 3.7]. Thus, the distribution of $\overline{X}$ does not approach the normal distribution for increasing n. The reason for this is, that the Cauchy distribution has neither a finite expectation nor a finite variance (and finite variance is essential for the central limit theorem, see also the paragraph following Theorem 5.5.14 in [12]).

This example is often called a pathological case not occurring in practice. On the other hand, the distribution a given sample comes from is usually not known, and in general there is no reason seen why instances

---

[8]The Cauchy distribution seems to have appeared first in the works of Pierre de Fermat in the midseventeenth century. See also Chapter 16 in [36] for a discussion of this distribution.

of a 'pathological' underlying distribution cannot occur. As will also be discussed in Sect. 8.3.2, the Cauchy distribution is the distribution of the ratio of two independent standard normal distributed random variables, so that 'taking ratios can lead to ill-behaved distributions' [12, p. 108].

2.6.3 Approximate Normality of Sample Statistics

The central limit theorem shows that the sample mean $\overline{X}$ has an approximate

$$N\left(\mu, \frac{\sigma^2}{n}\right)$$

distribution irrespective of the distribution the sample $X_1, \ldots, X_n$ comes from (provided the assumptions of the theorem are satisfied). Of course, when this distribution is known, then it may happen that the exact distribution of $\overline{X}$ for fixed sample size n is known. When the sample comes from a $N(\mu, \sigma^2)$ distribution, then the above approximate distribution is also the exact one.

Other sample statistics introduced above can also be shown to have an approximate normal distribution, at least under certain assumptions. From a practical point of view, we consider sample statistics only for fixed sample sizes, so that anyway the cdf of a sample statistic will reach normality only to a certain degree, thus justifying the use of the term 'approximate'. The order of magnitude of finite n, guaranteeing a sufficient degree of approximation is of course not the same for different sample statistics and does also depend on the actual distribution of the sample.

Approximate Normality of Sample Variance and Standard Deviation

When $X_1, \ldots, X_n$ is a sample from the $N(\mu, \sigma^2)$ distribution, the exact distribution of of the sample variance S^2 can be derived, see [54, p. 245]. For larger samples sizes this can be approximated by a normal distribution, see also [63, 6.3.6].

Theorem 2.9 Let $X_1, \ldots, X_n$ be a sample with expectation μ, finite variance σ^2 and finite fourth central moment μ_4.

(a) Let $G_n(x)$ denote the cdf of the random variable $\sqrt{n}(S^2 - \sigma^2)$. Then

$$\lim_{n\to\infty} G_n(x) = \Phi\left(\frac{x}{\sqrt{\mu_4 - \sigma^4}}\right).$$

(b) *Let $G_n(x)$ denote the cdf of the random variable $2\sigma\sqrt{n}(S - \sigma)$. Then*

$$\lim_{n\to\infty} G_n(x) = \Phi\left(\frac{x}{\sqrt{\mu_4 - \sigma^4}}\right).$$

From the above theorem, the sample variance S^2 is approximately

$$N\left(\sigma^2, \frac{\mu_4 - \sigma^4}{n}\right)$$

distributed. The sample standard deviation S is approximately

$$N\left(\sigma, \frac{\mu_4 - \sigma^4}{4\sigma^2 n}\right)$$

distributed.

Approximate Normality of Sample Skewness and Kurtosis

We have already introduced sample skewness $\sqrt{b_1}$ and kurtosis b_2 and noted that even under a sample from the $N(\mu, \sigma^2)$ distribution, the exact distributions of $\sqrt{b_1}$ and b_2 are not known. For larger n, however, these distributions are known to converge to normality.

Theorem 2.10 *Let $X_1, \ldots, X_n$ be a sample from the $N(\mu, \sigma^2)$ distribution.*

(a) *Let $G_n(x)$ denote the cdf of the random variable $\sqrt{b_1}$. Then*

$$\lim_{n\to\infty} G_n(x) = \Phi\left(\frac{x}{\sqrt{6/n}}\right).$$

(b) *Let $G_n(x)$ denote the cdf of the random variable b_2. Then*

$$\lim_{n\to\infty} G_n(x) = \Phi\left(\frac{x - 3}{\sqrt{24/n}}\right).$$

For a sample from the $N(\mu, \sigma^2)$ distribution, the sample skewness $\sqrt{b_1}$ is approximately

$$N(0, 6/n)$$

distributed, and the sample kurtosis b_2 is approximately

$$N(3, 24/n)$$

distributed. Especially for b_2, the sample size n must be quite large before the normal approximation appears to be of sufficient preciseness. Snedecor and Cochran [79, Sect. 3.13, 3.14] give $n = 150$ and $n = 1000$ as the respective minimal required sample sizes for the normal distribution approximation for $\sqrt{b_1}$ and b_2 to be of sufficient precision.

Approximate Normality of Sample Quantile

For a sample quantile $\widehat{\xi}_p$, the following is stated in [67, Sect. 6f.2].

Theorem 2.11 *Let $X_1, \ldots, X_n$ be a sample with strictly monotonic cdf $F(x)$ and continuous pdf $f(x)$. For $0 < p < 1$ let ξ_p denote the p-quantile, and let $f(\xi_p) > 0$. Let $G_n(x)$ denote the cdf of the random variable*

$$\frac{\sqrt{n}(\widehat{\xi}_p - \xi_p)}{\sqrt{p(1-p)}} \; ,$$

where $\widehat{\xi}_p$ is the sample p-quantile. Then $\lim_{n \to \infty} G_n(x) = \Phi(x f(\xi_p))$.

From the above theorem, the sample p-quantile $\widehat{\xi}_p$ is approximately

$$N\left(\xi_p, \frac{p(1-p)}{n f(\xi_p)^2}\right)$$

distributed. As a special case of this result we also obtain the approximate normality of the sample median by choosing $p = 0.5$.

When the sample comes from a $N(\mu, \sigma^2)$ distribution, then $\xi_{0.5} = \mu$ and $f(\xi_{0.5}) = 1/\sqrt{\sigma^2 2\pi}$. Hence, in that case, the sample median has an approximate normal distribution with mean μ and variance $\sigma^2 \pi/2n$.

2.7 Approximate Normality of the Binomial Distribution

Many statistical distributions (discrete and continuous) can be approximated via the normal distribution, see e.g. [63, Chapter 7]. In the following we only consider the approximation of the discrete binomial distribution.

2.7.1 The Binomial Distribution

The binomial distribution is a discrete distribution, naturally arising in relationship with a so-called *Bernoulli trial*. This is an experiment with exactly two possible outcomes, often called 'success' and 'failure'. More specifically, a Bernoulli trial satisfies the following requirements.

- The experiment consists of n identical trials with exactly one outcome out of two, the latter denoted by 'success' and 'failure'.
- The probability for 'success' remains the same from trial to trial and is denoted by p. (Then the probability for 'failure' remains the same from trial to trial, and is $1 - p$.)
- The trials are independent of each other.

Given these conditions, what is the probability $P(X = x)$ that the *number of 'successes'* X equals a given value x satisfying $0 \leq x \leq n$? It can be deduced from combinatoric considerations, see e.g. [12, Sect. 3.2], that

$$P(X = x) = \binom{n}{x} p^x (1-p)^{n-x}, \quad x = 0, 1, 2, \ldots, n .$$

The function

$$f(x) = \binom{n}{x} p^x (1-p)^{n-x} I_{\{0,1,2,\ldots,n\}}(x)$$

is a probability mass function (see Definition 1.3), and the corresponding distribution is called the *binomial distribution*.

A simple example for a Bernoulli trial is the repeated tossing of a coin with the two possible outcomes 'head' and 'tail'. If the probability for 'head' is known to be $p = 0.2$, then in an experiment with $n = 50$ repeated tosses, the probability for $x = 7$ 'heads' is

$$P(X = 7) = \binom{50}{7} 0.2^7 \, 0.8^{43} = 0.08701158 \, ,$$

and the probability for less than or equal to $x = 7$ 'heads' is

$$P(X \leq 7) = \sum_{x=0}^{7} \binom{50}{x} 0.2^x \, 0.8^{50-x} = 0.1904098 \, .$$

See also Fig. 2.10 for the probability mass function of the binomial distribution with $n = 50$ and $p = 0.2$.

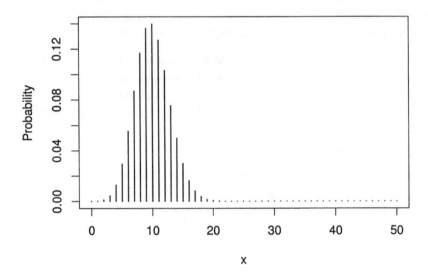

Fig. 2.10: Probability mass function of the binomial distribution with $n = 50$ and $p = 0.2$.

In real-life experiments, the assumptions of a Bernoulli trial are often at least approximately satisfied. If we want to know something about the number of persons in a population of size N having a 'yes' property out of 'yes' or 'no', then an actual random sample of a number n of persons admitting either 'yes' or 'no' can usually be seen as the outcome of a

Bernoulli trial, provided N is distinctly greater than n. (Often $n/N \leq 0.05$ is recommended. Otherwise the trials cannot be seen as being independent of each other.)

2.7.2 Approximation by the Normal Distribution

We have noted that the sum $\sum_{i=1}^{n} X_i$ of independent and identically distributed random variables $X_1, \ldots, X_n$ with arbitrary distribution can be approximated by a normal distribution. This property may be used to approximate the binomial by the normal distribution.

The de-Moivre-Laplace Theorem

In relationship with a Bernoulli trial we may define a random variable X_i as

$$X_i = \begin{cases} 1 \text{ if 'success' in trial } i \\ 0 \text{ if 'failure' in trial } i \end{cases},$$

in which case the number of 'successes' X is $X = \sum_{i=1}^{n} X_i$. Thus, a random variable X having a binomial distribution can always be seen as the sum of independent and identically distributed random variables, and therefore can be approximated by a normal distribution. It is easy to see that

$$\mathrm{E}(X_i) = p \quad \text{and} \quad \mathrm{Var}(X_i) = p(1-p) \,,$$

so that $X = \sum_{i=1}^{n} X_i$ has an approximate normal distribution with mean np and variance $np(1-p)$. As a rule of thumb for a situation when a 'good' approximation can be expected, sometimes the condition

$$\min\{np, n(1-p)\} \geq 5$$

is cited. The approximation of the distribution of X by the normal distribution may be used to compute probabilities concerning X.

Theorem 2.12 (de Moivre-Laplace Theorem) *If X has a binomial distribution, then*

$$\lim_{n \to \infty} P\left(a \leq \frac{X - np}{\sqrt{n}\sqrt{p(1-p)}} \leq b\right) = \Phi(b) - \Phi(a) \,,$$

is satisfied.

This theorem follows immediately from the central limit theorem. When we want to compute the probability for a binomial distributed random variable X to come out in $[c, d]$ for integers c and d with $0 \le c \le d \le n$, then from the above theorem

$$P(c \le X \le d) \approx \Phi\left(\frac{d - np}{\sqrt{np(1-p)}}\right) - \Phi\left(\frac{c - np}{\sqrt{np(1-p)}}\right).$$

Also,

$$P(X \le d) \approx \Phi\left(\frac{d - np}{\sqrt{np(1-p)}}\right) \quad \text{and} \quad P(X \ge c) \approx 1 - \Phi\left(\frac{c - np}{\sqrt{np(1-p)}}\right).$$

For example, if $n = 50$ and $p = 0.2$, then

$$P(X \le 7) \approx \Phi\left(\frac{7 - 10}{\sqrt{8}}\right) = \Phi(-1.060660) = 0.1444222.$$

Compared to the true probability of 0.1904098, this approximation is not very precise. It can, however, become better when applied with respect to a so called continuity correction.

Continuity Correction

When we approximate the discrete binomial by the continuous normal distribution, we apply two quite different types of distribution. Discrete distributions are concerned with probabilities for individual points while continuous distributions are concerned with probabilities for intervals.

In order to make the probability mass function of the binomial distribution 'more continuous', one may represent it by a pseudo-histogram in such a way that each possible outcome $x \in \{0, 1, 2, \ldots, n\}$ is the center of a bar with length one. Then, the probability $P(X \le x)$ is the area of all bars in the range from -0.5 to $x + 0.5$. Hence, an approximation

$$P(X \le x) \approx \Phi\left(\frac{x + 0.5 - np}{\sqrt{np(1-p)}}\right)$$

appears to be appropriate. This is called an approximation with *continuity correction*. See also Fig. 2.11 for an illustration with $n = 50$ and $p = 0.2$. The grey area corresponds to $\Phi[(x - np)/\sqrt{np(1-p)}]$, while the grey

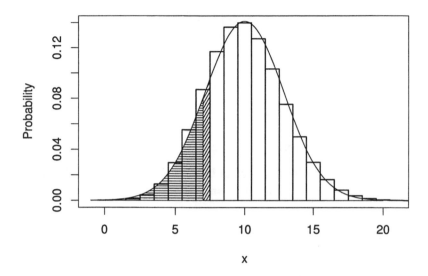

Fig. 2.11: Pseudo-histogram representation of the probability mass function of the binomial distribution with $n = 50$ and $p = 0.2$ for values $x = 0, \ldots, 21$, superimposed by the pdf of normal distribution with with mean np and variance $np(1 - p)$

plus shaded area corresponds to $\Phi[(x + 0.5 - np)/\sqrt{np(1 - p)}]$. As can be seen, the later is clearly a better approximation to the area of the bars in question. When we approximate $P(X \leq 7)$ for the case $n = 50$ and $p = 0.2$ by

$$\Phi\left(\frac{7.5 - 10}{\sqrt{8}}\right) = \Phi(-0.8838835) = 0.1883796\,,$$

this is seen to be closer to the true probability of 0.1904098 than the approximation without the continuity correction.

The continuity corrected version of the de Moivre-Laplace theorem yields

$$P(c \leq X \leq d) \approx \Phi\left(\frac{d + 0.5 - np}{\sqrt{np(1 - p)}}\right) - \Phi\left(\frac{c - 0.5 - np}{\sqrt{np(1 - p)}}\right)\,.$$

as well as

$$P(X \leq d) \approx \Phi\left(\frac{d + 0.5 - np}{\sqrt{np(1-p)}}\right)$$

and

$$P(X \geq c) \approx 1 - \Phi\left(\frac{c - 0.5 - np}{\sqrt{np(1-p)}}\right) ,$$

when X has a binomial distribution.

Further Approximation

An even better approximation to probabilities $P(X \leq x)$ for a binomial variable X using the functions $\Phi(\cdot)$ and $\phi(\cdot)$ is provided by the so-called *Gram-Charlier approximation*

$$P(X \leq x) \approx \Phi(z) - \frac{(1 - 2p)(z^2 - 1)\phi(z)}{6\sqrt{np(1-p)}}, \quad z = \frac{x + 0.5 - np}{\sqrt{np(1-p)}} ,$$

for $x \in \{0, 1, 2, \ldots, n\}$. For the case $n = 50$ and $p = 0.2$ this approximation gives $P(X \leq 7) = 0.1904673$, being very close to the true probability 0.1904098.

2.7.3   The Galton Board

The Galton board (also called *quincunx* by the inventor Galton [25]) may be used to demonstrate the occurrence of the normal pdf (actually the binomial pmf) shape in reality, see e.g. [39] for a review.

The board consists of n rows of nails, the bottom row meeting $n + 1$ slots which can be numbered by $0, 1, \ldots, n$. See Fig. 2.12 for a scheme when $n = 7$, the nails forming a *Pascal triangle*. A ball is dropped onto the nail in the top row, and takes its way through the rows of nails until it lands in one of the slots. In each row the ball hits one nail and can turn right with probability p and turn left with probability $1 - p$. In the end, the ball is in slot number i if it turned exactly i times to the right and $n - i$ times to the left.

By defining a 'right turn' as a 'success', this is a sequence of independent Bernoulli trials, where the chosen slot number i of the ball equals the number of successes, and thus can be seen as an outcome of the binomial distribution with parameters n and p.

If a considerable number of balls is brought through the board one after the other, the diagram formed by the heaps of balls in the slots will resemble the pmf of the binomial distribution, which in turn may be approximated by the pdf of the normal distribution, especially when $p = 1/2$, which is of course usually the considered case. The obtained diagram becomes more meaningful for a larger number of rows.

The actual production of a good quincunx is far from being easy, see also the comments in [39].

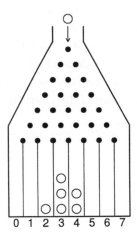

Fig. 2.12: Scheme of the Galton board with $n = 7$ rows of nails

2.8   Random Sample Generation

Statistical software packages usually come along with procedures to generate pseudo-random samples from well-known distributions in statistics. Such samples are quite useful in simulation studies when it is desired to know the underlying distribution of a sample, and then to study the behaviour of a specific statistical method. A description of methods for random sample generation is given e.g. in [70].

Usually the basis for the generation of a pseudo-random sample is a sequence of independent pseudo-random numbers, where a *pseudo-random number* is understood as the outcome of a random variable with uniform distribution on $(0, 1)$, having pdf $f(x) = I_{(0,1)}(x)$. Using a computer, such numbers are deterministically computed, but appear to be 'as if random'.

To generate a pseudo-random sample from the normal distribution with given parameters μ and σ^2, it is only necessary to obtain a sample $z_1, \ldots, z_n$ from the standard normal distribution, since then the transformed values $x_i = \mu + \sigma z_i$ constitute the required sample.

In principle, there are several possibilities to generate a standard normal pseudo-random sample, the most popular being the *Box-Muller* [9] *method*. The corresponding algorithm is based on the fact, that if u_1 and u_2 are two observations taken independently from the uniform distribution on $(0, 1)$, then z_1 and z_2 can be seen as two observations taken independently from the standard normal distribution, where

$$z_1 = r \cos(\theta) \quad \text{and} \quad z_2 = r \sin(\theta) \, ,$$

with

$$r = \sqrt{-2 \ln(u_1)} \quad \text{and} \quad \theta = 2\pi u_2 \, .$$

This can easily be applied to obtain a sample of arbitrary size n by independently repeating $n/2$ times[9] the generation of values u_1 and u_2.

A variant of the Box-Muller method is the so-called *polar method*, see e.g. [70, Sect. 5.3], avoiding the direct computation of sine and cosine, the latter being computationally not very efficient. The starting point is the same as above, namely the generation of two observations u_1 and u_2 taken independently from the uniform distribution on $(0, 1)$. Then let

$$v_1 = 2u_1 - 1, \quad v_2 = 2u_2 - 1, \quad \text{and} \quad s = v_1^2 + v_2^2 \, .$$

If $s > 1$, two new observations u_1 and u_2 must be generated. Otherwise, z_1 and z_2 can be seen as two observations taken independently from the standard normal distribution, where

---

[9] $(n+1)/2$ times if n is odd, in which case one superfluous observation is generated.

$$z_1 = \sqrt{\frac{-2\ln(s)}{s}} v_1 \quad \text{and} \quad z_2 = \sqrt{\frac{-2\ln(s)}{s}} v_2 \ .$$

As noted in [70, p. 75], this procedure will on average require the generation of 2.546 uniform random numbers, 1 logarithm, 1 square root, 1 division, and 4.546 multiplications to generate two independent observations from the standard normal distribution.

Chapter 3

Checking for Normality

When observations $x_1, \ldots, x_n$ of a variable X are given and may be assumed to come from a random sample, then it is often of interest to find out whether the underlying distribution can be normal, or whether there are specific features invalidating a possible normality assumption. Tools for checking for normality can be the computation of specific *sample characteristic values*, specific *graphical representations* of the data, and the application of statistical *significance tests*, the latter being discussed in the subsequent chapter. From a data analytical point of view, the combination of available methods is advisable. Nonetheless, for checking for normality solely from given data points (without having some knowledge about a possible normal distribution of the corresponding random variable X), one can in general not do better than the data permits. It may happen, that the data comes from a normal distribution but looks rather non-normal, and it may also happen that the given data looks quite normal, while in fact coming from a different distribution.

3.1 Sample Characteristic Values

The $N(\mu, \sigma^2)$ distribution has a number of characteristics values already discussed in the previous chapter, e.g. the interquartile range divided by the standard deviation is $2u_{0.75} \approx 1.34$.

In order to check whether a sample value $x_1, \ldots, x_n$ can come from a normal distribution, one may compute the values of the corresponding sample characteristics and see whether there are strong deviations from the hypothetical values (under normality) or not. A list of such characteristics is shortly discussed in the following.

- The difference taken between the arithmetic mean $\bar{x} = \frac{1}{n} \sum_{i=1}^{n} x_i$ and the value m of the sample median M and then divided by the standard deviation $s = \sqrt{s^2}$, where

$$s^2 = \frac{1}{n-1} \sum_{i=1}^{n} (x_i - \bar{x})^2 \, ,$$

  should be near zero when the distribution is symmetric. A negative value can indicate skewness to the left, a positive value can indicate skewness to the right.
- When we denote the value of the sample IQR by iqr, then the quotient iqr/s should be near 1.34 under normality.
- When similarly mad denotes the value of the sample MAD, then mad/s should be near 0.67 under normality.
- An empirical k standard deviation interval can be defined as

$$(\bar{x} - ks, \bar{x} + ks) \, .$$

For $k = 1, 2, 3$ we can compute the proportion of the data falling in the respective interval, and compare it with the probability of the corresponding $k\sigma$ interval under normality.

- The values $m_j = \frac{1}{n} \sum_{i=1}^{n} (x_i - \bar{x})^j$ can be used to compute sample skewness $\sqrt{b_1} = m_3/m_2^{3/2}$ and sample kurtosis $b_2 = m_4/m_2^2$. A strong deviation of $\sqrt{b_1}$ from zero may also indicate non-symmetry. A deviation of b_2 from 3 may reveal a more peaked or a flatter distribution than can be expected under normality.

Table 3.1 shows sample characteristics together with their hypothetical values under normality. The values $(\bar{x} - m)/s$ and $\sqrt{b_1}$ are not specifically related to the normal distribution, but to any symmetric distribution. The other sample characteristics can be seen to be more specific to the normal distribution, although it might be possible to find non-normal distributions having one or the other characteristic identical to the normal distribution.

sample characteristic	value under normality
$(\overline{x} - m)/s$	0
iqr$/s$	1.34
mad$/s$	0.67
$\#\{x_i : x_i \in (\overline{x} - s, \overline{x} + s)\}/n$	0.68
$\#\{x_i : x_i \in (\overline{x} - 2s, \overline{x} + 2s)\}/n$	0.95
$\#\{x_i : x_i \in (\overline{x} - 3s, \overline{x} + 3s)\}/n$	1.00
$\sqrt{b_1}$	0
b_2	3

Table 3.1: Sample characteristics and corresponding hypothetical values

Hence, the above characteristics should be regarded together. In addition, it should be kept into mind that even when the sample comes from a normal distribution, a sample characteristic will only be somewhat near the corresponding hypothetical value. It is therefore not easy to asses a sample characteristic as being 'too far away' from the hypothetical value to violate normality. Sample skewness and kurtosis are also used as test statistics, allowing for the determination of critical values for assessing a too strong deviation, see [86].

$(\overline{x} - m)/s$	iqr$/s$	mad$/s$	$\#\{1s\}/n$	$\#\{2s\}/n$	$\#\{3s\}/n$	$\sqrt{b_1}$	b_2
-0.0648	1.3147	0.6367	0.69	0.95	1.00	-0.1722	3.1335
(0.00)	(1.34)	(0.67)	(0.68)	(0.95)	(1.00)	(0.00)	(3.00)

Table 3.2: Sample characteristics for the female athletes weights data from Table 1.1 (normal hypothetical values in brackets)

Example 3.1 For the data of $n = 100$ female athletes weights from Table 1.1, we obtain the sample characteristic values presented in Table 3.2. The computed values look quite in accordance with the normal distribution.□

Recommendations

Sample characteristics should *never* be looked at separately, but *always* be accompanied by a graphical representation of the empirical distribution of the data via a histogram and/or a boxplot.

3.2 Graphics

In addition to the computation of individual sample characteristics, one may consider the empirical distribution of the data as a whole, and try to verify whether it can be the empirical version of a normal distribution or not. To achieve this, a graphical representation of the data appears to be appropriate. Since we usually have no knowledge about the parameters μ and σ^2 of the possible $N(\mu, \sigma^2)$ distribution, these are replaced by estimates $\widehat{\mu}$ and $\widehat{\sigma}^2$ if necessary. Three important tools are:

- The histogram (empirical density). The histogram is plotted and superimposed by the probability density function (pdf) of the $N(\widehat{\mu}, \widehat{\sigma}^2)$ distribution.
- The empirical cumulative distribution function (ecdf). The ecdf is plotted together with the cumulative distribution function of the $N(\widehat{\mu}, \widehat{\sigma}^2)$ distribution.
- Empirical quantiles. The sample order statistics are plotted against expected ordered quantiles of the standard normal distribution (*normal quantile-quantile plot*).

The third procedure is the most common and most effective one.

3.2.1 The Histogram

We have already considered the histogram in Sect. 1.1.3 to display the frequency distribution of a variable X with observations $x_1, \ldots, x_n$. In order to get some idea whether the observations can come from a normal distribution, we may estimate μ and σ^2 by

$$\widehat{\mu} = \overline{x} = \frac{1}{n} \sum_{i=1}^{n} x_i \quad \text{and} \quad \widehat{\sigma}^2 = s^2 = \frac{1}{n-1} \sum_{i=1}^{n} (x_i - \widehat{\mu})^2,$$

and compare the density of $N(\bar{x}, s^2)$ with the histogram. Fig. 3.1 (a) shows histograms and estimated pdfs for the female athletes weights data ($n = 100$) from Table 1.1 and for the leukemia latency period data ($n = 20$) from Table 3.3. While the normal density seems to fit the athletes weights data quite well, this appears to be not the case for the leukemia latency data[1].

	latency period				
1 – 5	16	72	54	52	62
6 – 10	12	21	44	56	32
11 – 15	60	60	168	66	50
16 – 20	11	132	48	120	72

Table 3.3: Induced acute leukemia latency period following chemotherapy as given in [86, Data Set 9]

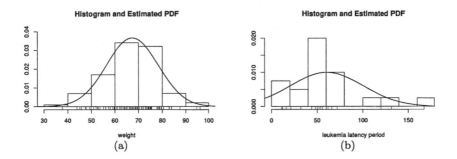

(a) (b)

Fig. 3.1: Female athletes weights (a) from Table 1.1 and leucemia latency periods (b) from Table 3.3

---

[1]Of course, conclusions with respect to the leukemia latency data must be regarded with caution in view of the rather small sample size $n = 20$.

Number of Classes

As already noted, the visual impression of the histogram depends on the choice of classes. For classes of equal length, there exist different proposals for the choice of number of classes, including Sturges' formula already mentioned in Sect. 1.1.3. Scott [74] obtains an asymptotically optimal choice[2] for the width h of the classes, given as

$$h = \left(\frac{6}{\int_{-\infty}^{\infty} [f'(x)]^2 \, dx} \right)^{1/3} n^{-1/3} \, ,$$

where $f(x)$ is the pdf the sample comes from. As $f(x)$ will not be known, Scott proposes to take the normal pdf as a reference. Then $\int_{-\infty}^{\infty} [f'(x)]^2 \, dx = 1/(4\sqrt{\pi}\sigma^3)$ and $h = (24\sqrt{\pi}\sigma^3)^{1/3} n^{-1/3}$. Replacing σ by the actual sample standard deviation s yields a class width of

$$h = 3.49 s n^{-1/3} \, .$$

From this, the number of classes according to *Scott's formula* is the smallest integer not less than

$$(x_{(n)} - x_{(1)})/h \, ,$$

where $x_{(1)}$ is the smallest and $x_{(n)}$ is the largest observation. As noted by Scott this choice will not necessarily result in a normal looking histogram. A further alternative for the choice of class width is the proposal by Freedman and Diaconis [24],

$$h = 2 \, \text{iqr} \, n^{-1/3} \, ,$$

where iqr is the actual interquartile range of the data. Again, the corresponding number of classes is then the smallest integer not less than $(x_{(n)} - x_{(1)})/h$.

Recommendations

A histogram together with the estimated normal pdf can be an additional help for finding out, whether the data can be assumed to come from a normal distribution or not. It should, however, not be the primary tool of

---

[2]With respect to the so-called integrated mean squared error, see [74] for details.

analysis. The visual impression of the histogram depends on the classes, and the addition of the estimated normal pdf is rather suggestive. Especially when the data is symmetrically distributed, the eye tends to polish off and ignore possible discrepancies.

3.2.2 The ECDF

Let $X_1, \ldots, X_n$ be a sample from a distribution with cumulative distribution function $F(x)$ (also called *parent cdf*), and let $x_1, \ldots, x_n$ be the corresponding observations.

Empirical Cumulative Distribution Function

The *empirical cumulative distribution function* (ecdf) is a step function

$$F_n(x) = \frac{\text{number of observations} \leq x}{n}, \quad x \in \mathbb{R},$$

recording the proportion of observations less than or equal to x. More precisely,

$$F_n(x) = \begin{cases} 0 & x < x_{(1)} \\ i/n & x_{(i)} \leq x < x_{(i+1)} \\ 1 & x_{(n)} \leq x \end{cases},$$

where $x_{(1)}, \ldots, x_{(n)}$ are the ordered observations. For any fixed x, the ecdf $F_n(x)$ is a random variable with

$$\mathrm{E}[F_n(x)] = F(x) \quad \text{and} \quad \mathrm{Var}[F_n(x)] = \frac{1}{n} F(x)(1 - F(x)).$$

This implies that for any fixed x, the ecdf $F_n(x)$ is a consistent estimator for the parent cdf $F(x)$. In addition, the *Theorem by Glivenko and Cantelli*

$$P\left[\lim_{n \to \infty} \sup_{-\infty < x < \infty} : F_n(x) - F(x)| = 0\right] = 1$$

holds true, showing that with probability one, the convergence of $F_n(x)$ to $F(x)$ is uniform in x, see [54, p. 507].

Comparison of ECDF and Estimated Normal CDF

For the purpose of finding out, whether the parent cdf $F(x)$ is the $N(\mu, \sigma^2)$ cumulative distribution function

$$F^*(x) := \Phi\left(\frac{x - \mu}{\sigma}\right) ,$$

we may display the functions $F_n(x)$ and $F^*(x)$ together in one plot. If the parent cdf $F(x)$ equals $F^*(x)$, the ecdf $F_n(x)$ should be rather similar to $F^*(x)$, while, if the parent cdf $F(x)$ is not $F^*(x)$, some differences between $F_n(x)$ and $F^*(x)$ should become apparent. Of course, we still assume μ and σ^2 to be unknown, so that again for computing $F^*(x)$ these must be replaced by the estimates $\widehat{\mu} = \overline{x}$ and $\widehat{\sigma}^2 = s^2$.

Figures 3.2 (a) and 3.2 (b) show the ecdf and the estimated normal cumulative distribution function for the data sets from Table 1.1 and Table 3.3, respectively.

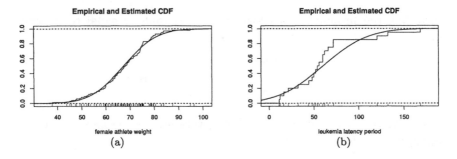

Fig. 3.2: Female athletes weights (a) from Table 1.1 and leucemia latency periods (b) from Table 3.3

Recommendations

As the histogram, the ecdf plot can help to find out, whether the data can assumed to be normally distributed or not, but should not be the primary tool of analysis. Especially for (almost) symmetrically distributed data it can be difficult to figure out possible discrepancies between the ecdf and the normal cumulative distribution function.

3.2.3 The Normal Quantile-Quantile Plot

Let $X_1, \ldots, X_n$ be a random sample and let $X_{(1)}, \ldots, X_{(n)}$ be the corresponding order statistics with given observations $x_{(1)}, \ldots, x_{(n)}$.

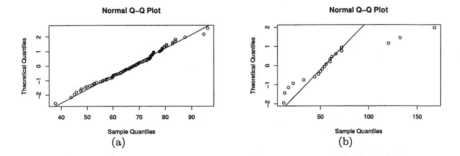

Fig. 3.3: Female athletes weights (a) from Table 1.1 and leucemia latency periods (b) from Table 3.3

Observed Order Statistics

Let us assume for the moment that the sample comes from the $N(\mu, \sigma^2)$ distribution. Moreover, let $Z_1, \ldots, Z_n$ be a random sample from the $N(0, 1)$ distribution and let $Z_{(1)}, \ldots, Z_{(n)}$ be the corresponding order statistics with expectations $m_1, \ldots, m_n$, $m_i = E(Z_{(i)})$. In view of the identity

$$E(X_{(i)}) = \mu + \sigma m_i \,,$$

for the observation $x_{(i)}$, the approximate identity

$$x_{(i)} \approx \mu + \sigma m_i,$$

should hold. This means, that, without knowing μ and σ, it can nonetheless be assumed that the data points $(m_i, x_{(i)})$ are nearly on a straight line when the observations come from a normal distribution.

For arbitrary sample size n, the m_i can only be computed numerically, but according to Blom [8] may be approximated by $m_i = \Phi^{-1}(p_i)$, where

$$p_i = \frac{i - \delta}{n - 2\delta + 1}$$

for some $0 \le \delta < 1$. The p_i are also called *probability points*. The most common choices are $\delta = 3/8$ and $\delta = 1/2$, giving

$$m_i = \Phi^{-1}\left(\frac{i - 3/8}{n + 1/4}\right) \quad \text{and} \quad m_i = \Phi^{-1}\left(\frac{i - 1/2}{n}\right),$$

respectively.

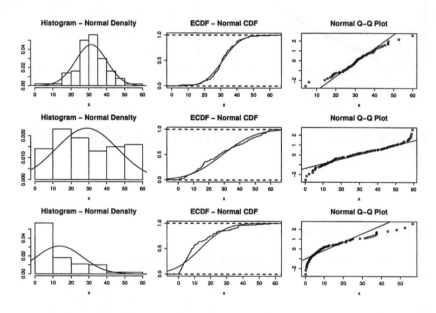

Fig. 3.4: Histogram, ecdf and normal Q-Q plot for three non-normal samples of size $n = 100$

Normal Quantile-Quantile Plot

Now, a *normal quantile-quantile plot* (normal Q-Q plot) is simply a plot of the points $(x_{(i)}, m_i)$ (or $(m_i, x_{(i)})$), where the m_i are obtained by the above approximation with $\delta = 1/2$. (If $n \leq 10$, then one may prefer $\delta = 3/8$.) If the observations come from a normal distribution, then, as reasoned above, the points $(x_{(i)}, m_i)$ should be nearly on a straight line.

In order to get a better impression about the possible deviation from linearity, a straight line through the points $(\tilde{x}_{0.25}, \Phi^{-1}(0.25))$ and $(\tilde{x}_{0.75}, \Phi^{-1}(0.75))$ can be added, where $\tilde{x}_{0.25}$ and $\tilde{x}_{0.75}$ stand for lower and upper quartile of the data, respectively. Figures 3.3 (a) and 3.3 (b) show the normal Q-Q plots for the data sets from Table 1.1 and Table 3.3, respectively.

Figure 3.4 shows a comparison of the histogram, the ecdf and the normal Q-Q plot for three non-normal samples each of size $n = 100$. The first row in Fig. 3.4 is created from a symmetric distribution (precisely: Laplace distribution), having heavier tails than the normal distribution. The Q-Q plot shows a linear behaviour in the centre of the data, but some essential deviations at both ends. The second row is created from a distribution with no tails (precisely: uniform distribution). The Q-Q plot reveals an S-shaped pattern. The third row is created from a right-skewed distribution (precisely: exponential distribution). The Q-Q plot shows a curved pattern.

Normal Probability Paper

When no computer is available to produce a normal Q-Q plot, this may also be done by hand on *normal probability paper*. Such paper is simply a printed xy-coordinate plane, where probabilities p (usually between 0.001 and 0.999) are marked at positions $\Phi^{-1}(p)$ along the y-axis. The x-axis can be labeled by the user and will comprise the range of the given observations. See Fig. 3.5 (a) for a scheme. Minimum and maximum on the x-axis must not necessarily be identical to $x_{(1)}$ and $x_{(n)}$, but should of course be smaller and greater, respectively.

To check for normality, the points $(x_{(i)}, p_i)$ for $p_i = (i - 0.5)/n$ are drawn on normal probability paper. Similarly to the normal quantile quantile plot, these points should nearly lie on a straight line to indicate normality. See Fig. 3.5 (b) for a plot of the female athletes weights data from Table 1.1.

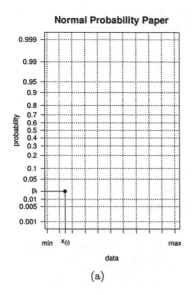

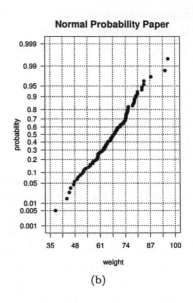

(a)   (b)

Fig. 3.5: Scheme of normal probability paper (a) and with plotted points for the female athletes weights data (b)

Recommendations

A normal Q-Q plot is not easy to interpret, since even if the data does in fact come from a normal distribution, there may occur random deviations from the ideal straight line. Nonetheless, deviations from normality are easier to make out as from an ecdf plot (together with the estimated normal cdf) and do not depend on classes as the histogram. The normal Q-Q plot does not even require estimates for μ and σ^2. When it comes to the question whether the 'straight line property' is violated or not, one may also compute the squared empirical Pearson correlation coefficient between ordered sample values and expected quantiles as an additional aid. Then, it can be assessed by the Shapiro-Francia test procedure (see Sect. 4.4.1) whether the correlation is too low to justify the normality assumption.

3.3 Summary

In order to check for normality, sample characteristics may be computed and compared with corresponding hypothetical values under normality. Strong deviations may be seen as an indication for non-normality. Such analysis should *always* be accompanied by a graphical inspection of the data, the normal quantile-quantile plot being the most common graphical device for this purpose. In addition to the graphical methods presented in the previous section, others may also applied. For example a *boxplot* may reveal skewness or the presence of outliers in a data set, not being in accordance with normality.

Chapter 4

Testing for Normality

We have already noted in Sect. 3.1 that we may consider specific sample characteristics whose actual values are meant to hint to possible violations of normality. The problem with these values is that we often cannot say what 'too small' or 'too large' is. Statistical significance tests are designed to overcome this problem. Here, there are characteristics (test statistics) whose distribution is known under the hypothesis of normality. Then a computed value of the test statistic is 'too small' or 'too large' when its occurrence can be assessed as being 'quite unlikely'. This can further be narrowed with the help of a critical value. A comprehensive review of testing for normality is given in [86].

4.1 General Remarks

Before discussing some test procedures, the principles of significance tests are shortly presented in Sect. 4.1.1, and a classification of existing goodness-of-fit tests is given in Sect. 4.1.2.

4.1.1 Significance Tests

In the following we will discuss testing the hypothesis

H_0: a random sample $X_1, \ldots, X_n$ comes from the normal distribution $N(\mu, \sigma^2)$ with μ and σ^2 unknown versus the alternative hypothesis

H_1: not H_0 .

Since the hypothesis H_0 only specifies a set of possible distributions

$$\{N(\mu, \sigma^2) : \mu \in \mathbb{R}, \sigma^2 > 0\} ,$$

it is called a *composite* hypothesis, as opposed to a *simple* hypothesis in which the distribution is completely specified.

Clearly, the alternative hypothesis H_1 does not contain a specific alternative distribution. Any test designed to handle such a problem is called an *omnibus test*.

The Test Procedure

For testing H_0, a statistic $T = T(X_1, \ldots, X_n)$ is considered, being designed in such a way that certain values of T are unlikely to occur under the hypothesis H_0, but more likely to occur under the alternative hypothesis H_1. Then, if one of these certain values of T is actually observed, the hypothesis H_0 is rejected.

For testing the hypothesis H_0 of normality, usually nonnegative test statistics T are used, leading to rejection of H_0 when T is too large. Now, what does 'too large' actually mean? To find an answer, we may proceed in three steps:

Step 1. Fix a rather small value $\alpha \in (0, 1)$, usually $\alpha = 0.01$ or $\alpha = 0.05$.

Step 2. Find the *critical value* c such that the probability for T to be greater than c does not exceed α when H_0 is true, i.e. find c with

$$P_{H_0}(T > c) \leq \alpha .$$

The critical value actually depends on the choice of α in Step 1, so that we may write $c(\alpha)$ instead of c. When the hypothesis H_0 is true, then it is rather unlikely that T takes a value greater than $c(\alpha)$, but if so, this will lead to rejection of H_0, see the next step.

Step 3. Let $t = T(x_1, \ldots, x_n)$ be the actual value of $T = T(X_1, \ldots, X_n)$. Reject H_0 at level α if $t > c(\alpha)$. Otherwise do not reject H_0.

The p-value

The above describes a (right-tailed) level-α-test procedure. As an extension to this method, one may also compute the probability

$$p_* = \mathrm{P}_{H_0}(T > t) \,,$$

being called the *p-value* (probability value) of the test. The p-value of a test is usually computed by statistical software packages and can be used in two ways. First, it may be employed to alter Step 3 in the above procedure as

Step 3. Reject H_0 at level α if $p_* \leq \alpha$. Otherwise do not reject H_0.

Of course, as soon as p_* is known it is possible to fix α such that rejection of H_0 is guaranteed, but this is *not* a proper test procedure then.

Second, as an alternative to the strict rejection/non-rejection test procedure, one may regard the p-value as a measure for the *degree of support* of H_0. The smaller the p-value, the less H_0 is supported.

The Power

The above test procedure ensures that the probability for rejecting H_0 when H_0 is in fact true cannot exceed α, i.e.

$$\mathrm{P}_{H_0}(\text{reject } H_0) \leq \alpha \,.$$

Rejecting H_0 when H_0 is in fact true is known as the *Type I error*. When H_0 is accepted although H_1 is true, this is called *Type II error*. Naturally, the probability for the Type II error $\mathrm{P}_{H_1}(\text{accept } H_0)$ should also be small, and when several level-α-tests are available, it is aimed to find the one with the smallest Type II error. This is equivalent to finding a level-α-test with largest power, where the *power* is defined to be

$$\mathrm{P}_{H_1}(\text{reject } H_0) \,.$$

The determination of the power of a test is usually not easy. In addition, for an omnibus test no specific alternative to the normal distribution is given, and it will not be possible to find a most powerful test with regard to all possible alternatives. Hence, the power of a test is often studied via simulations for a wide range of non-normal alternatives and a variety of different sample sizes, see also [1, Sect. 9.4, 9.5] for a review.

When a test is known to have a low power, then we are rather tentative in accepting H_0, since the probability for the (Type II) error is quite high. Hence, a non-rejection result obtained from such a weak power test procedure only little helps in our decision. This is also true with regard to the p-value. If a computed p-value supports H_0 to a certain extent, but the corresponding test has weak power, it remains rather uncertain whether H_0 can in fact assumed to be true.

4.1.2 Classification of Tests

There exist various goodness-of-fit omnibus tests which may be applied for testing for normality. Mainly, these tests fall into three categories, being related to the graphical methods described in Sections 3.2.1, 3.2.2, and 3.2.3:

- Tests of chi-squared type. These tests require the building of classes from the full sample. The counted number of observations in each class is compared to the expected number of observations, where the expectation is taken with respect to the hypothetical distribution. If the difference (measured by the test statistic) is too large, the hypothesis of normality is rejected. Pearson's classical *chi-square test* is discussed in Sect. 4.2.

- Tests based on the empirical cumulative distribution function. These tests are based on measures of the difference between empirical and hypothetical cumulative distribution function. If both functions are too distinct, the hypothesis of normality is rejected. The *Lilliefors (Kolmogorov-Smirnov) test*, the *Cramér-von Mises test*, and the *Anderson-Darling test* are discussed in Sect. 4.3.

- Tests based on regression and correlation. Essentially, the test statistics are measures of the strongness of the linear relationship between the sample ordered values and the expected values of standard normal order statistics. If the linear relationship is not strong enough, the hypothesis of normality is rejected. The *Shapiro-Francia test* and the *Shapiro-Wilk test* are discussed in Sect. 4.4.

The chi-square test from Pearson and the Lilliefors (Kolmogorov-Smirnov) test are the most famous and widely applied omnibus tests for normality. Nonetheless, both seem to be inferior to the Cramér-von Mises test, the Anderson-Darling test, the Shapiro-Francia test and the Shapiro-Wilk test, and can thus not be recommended for testing for normality.

4.2 The Chi-Square Test

The chi-square test of normality from Pearson [65] is related to the graphical procedure described in Sect. 3.2.1. The real line $(-\infty, +\infty)$ is divided into k classes

$$(-\infty, \xi_1], (\xi_1, \xi_2], (\xi_2, \xi_3], \ldots, (\xi_{k-2}, \xi_{k-1}], (\xi_{k-1}, +\infty) .$$

Then the number n_i, $i = 1, \ldots, k$, of observations in each class is determined and compared to the number of observations $\tilde{n}_i$ one would expect if the data points were generated from a $N(\hat{\mu}, \hat{\sigma}^2)$ distribution, where

$$\hat{\mu} = \overline{x} \quad \text{and} \quad \hat{\sigma}^2 = s^2 = \frac{1}{n-1} \sum_{i=1}^{n} (x_i - \overline{x})^2 .$$

The expected numbers are computed as

$$\tilde{n}_i = n \left(\Phi \left(\frac{\xi_i - \hat{\mu}}{\hat{\sigma}} \right) - \Phi \left(\frac{\xi_{i-1} - \hat{\mu}}{\hat{\sigma}} \right) \right) ,$$

where $\xi_0 := -\infty$ so that $\Phi(\frac{\xi_0 - \hat{\mu}}{\hat{\sigma}}) = 0$, and $\xi_k := +\infty$ so that $\Phi(\frac{\xi_k - \hat{\mu}}{\hat{\sigma}}) = 1$. Note that the $\tilde{n}_i$ are not necessarily integers. This is, however, in accordance with the fact that the expectation of a random variable taking only integer values must not necessarily be an integer.

The test statistic measuring the difference between the counted numbers n_i and the expected numbers $\tilde{n}_i$ of observations is given as

$$X^2 = \sum_{i=1}^{k} \frac{(n_i - \tilde{n}_i)^2}{\tilde{n}_i} .$$

The hypothesis H_0 of normality is rejected if the computed value X is too large. We will come a little later to the question what actually 'too large' means.

Choice of Classes

For now, note that the described procedure for obtaining X is not unique, since it depends on the number k of classes and on the choice of the ξ_i. For a histogram, usually the classes are formed to have equal length. For the chi-square test, it is proposed to choose *equiprobable* classes. This means that the classes are chosen such that the probability for a normally distributed random variable with expectation $\hat{\mu}$ and variance $\hat{\sigma}^2$ to take a value in a specific class is the same for each class. Hence, the ξ_i are simply the (i/k)-quantiles of the $N(\hat{\mu}, \hat{\sigma}^2)$ distribution, i.e. ξ_i satisfies $\Phi(\frac{\xi_i - \hat{\mu}}{\hat{\sigma}}) = i/k$ for $i = 1, \dots, k-1$. Then, the expected numbers $\tilde{n}_i$ are all identical to n/k and the test statistic X may also be written as

$$X^2 = \frac{k}{n} \sum_{i=1}^{k} n_i^2 - n .$$

Next, an adequate choice of the number of classes k matters. Without further commenting on it, we adopt a proposal by Moore [55, p. 70] and choose k as the smallest integer not less than $2n^{(2/5)}$.

Critical Values

Let us now turn to the question how to assess X as being 'too large' to justify the hypothesis of normality. It is common practice to compute critical values (or p-values) from a χ^2_{k-1-m} distribution for $m = 2$. This is motivated by the fact that in case of a simple hypothesis (when μ and σ^2 are known), the statistic X^2 is asymptotically χ^2_{k-1} distributed, and in order to account for the estimation of the two parameters μ and σ^2, $m = 2$ further degrees of freedom are subtracted. It should, however, be noted that this is *not* a correct procedure when μ and σ^2 are estimated by $\bar{x}$ and s^2 as above, see e.g. [55, Sect. 3.2] or [86, Sect. 5.2] for a discussion. All what can be said is that the critical values fall between those of χ^2_{k-1-2} and those of χ^2_{k-1}, see [55, p. 68].

Example 4.1 For the data of $n = 100$ female athletes weights from Table 1.1, equiprobable classes (under the hypothesis of normality) together with the number of counted and expected observations are given in Table 4.1. A graphical illustration is given in Fig. 4.1

class i	interval of weights	n_i	$\tilde{n}_i$
1	$(-\infty, 51.77623]$	9	7.69231
2	$(51.77623, 56.20791]$	7	7.69231
3	$(56.20791, 59.30528]$	4	7.69231
4	$(59.30528, 61.85856]$	8	7.69231
5	$(61.85856, 64.14011]$	9	7.69231
6	$(64.14011, 66.28852]$	5	7.69231
7	$(66.28852, 68.39648]$	9	7.69231
8	$(68.39648, 70.54489]$	9	7.69231
9	$(70.54489, 72.82644]$	4	7.69231
10	$(72.82644, 75.37972]$	16	7.69231
11	$(75.37972, 78.47709]$	5	7.69231
12	$(78.47709, 82.90877]$	10	7.69231
13	$(82.90877, +\infty)$	5	7.69231
Totals		100	100

Table 4.1: Weights of 100 female athletes

Since $2n^{(2/5)} = 12.61915$, the number of classes is $k = 13$. The ξ_i are computed as the $(i/13)$-quantiles from the $N(\bar{x}, s^2)$ distribution, $i = 1, \ldots, 12$.

Now, the chi-square test statistic is computed as $X = 17$. The p-value obtained from the χ^2_{k-3} distribution is $p_* = 0.07436$, while obtained from the χ^2_{k-1} distribution it is $p_* = 0.1496$.

In order to demonstrate the dependence of the chi-square test on the classes, let us now decide to choose $k = 12$, but otherwise proceed as above. Then we obtain $X = 5.12$ and the p-value obtained from the χ^2_{k-3} distribution is $p_* = 0.8237$, while obtained from the χ^2_{k-1} distribution it is $p_* = 0.9252$. $\qquad \square$

Recommendations

The chi-square test is usually *not* recommended for testing the hypothesis of normality. The test result depends on the choice of classes and compared to other tests, the chi-square test has low power.

The chi-square test may be used when the full sample is not available, but the data is already divided into classes. In that case we do not have to bother with the choices of k and ξ_i, $i = 1, \ldots, k-1$. (Of course we do not necessarily have equiprobable classes then.) If the original number n of observations and the number k of classes is not small, we may obtain reasonable estimates $\widehat{\mu}$ and $\widehat{\sigma}^2$ from the classes and may compute critical values (or p-values) from the χ^2_{k-3} distribution.

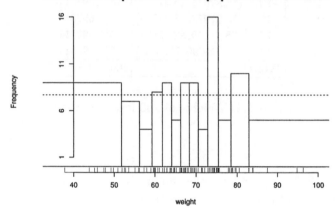

Fig. 4.1: Absolute frequencies for the equiprobable classes from Table 4.1

4.3   Tests Based on the ECDF

There exist a number of tests for normality, based on a comparison of the ecdf $F_n(x)$ with the cumulative distribution function $F(x)$ of the $N(\mu, \sigma^2)$ distribution. Since only the composite hypothesis is considered here, again μ and σ^2 are replaced by

$$\widehat{\mu} = \overline{x} \quad \text{and} \quad \widehat{\sigma}^2 = s^2 = \frac{1}{n-1} \sum_{i=1}^{n} (x_i - \overline{x})^2 .$$

This is related to the graphical comparison in Sect. 3.2.2 with the distinction that a significance test is based on some measure for the difference of the two functions, this measure being the test statistic. The hypothesis of normality is then rejected when the measured difference is too large.

For the composite hypothesis it is usually difficult to say what 'too large' actually means, since the distribution of the test statistic won't be fully known, so that critical values or p-values are difficult to determine. Often, there exist approximations based on simulation studies.

The following three *ecdf tests* (Lilliefors , Cramér-von Mises, Anderson-Darling) require the computation of the values

$$p_{(i)} = \Phi\left(\frac{x_{(i)} - \widehat{\mu}}{\widehat{\sigma}}\right),$$

where $x_{(1)}, \ldots, x_{(n)}$ are the given observations $x_1, \ldots, x_n$ arranged in ascending order.

4.3.1 The Lilliefors (Kolmogorov-Smirnov) Test

The Kolmogorov-Smirnov test for normality considers the maximal absolute difference

$$D = \max_x |F_n(x) - F(x)|$$

between the empirical cumulative distribution function $F_n(x)$ and the cumulative distribution function $F(x)$ of the $N(\mu, \sigma^2)$ distribution. It can also be written as $D = \max\{D^+, D^-\}$, where

$$D^+ = \max_x\{F_n(x) - F(x)\} \quad \text{and} \quad D^- = \max_x\{F(x) - F_n(x)\}.$$

If an actual value of D is too large, the hypothesis of normality is rejected.

For the composite hypothesis of normality, the parameters μ and σ^2 are not known, but replaced by estimates $\widehat{\mu} = \bar{x}$ and $\widehat{\sigma}^2 = s^2$. Then, for testing H_0 compute $D = \max\{D^+, D^-\}$, where

$$D^+ = \max_{i=1,\ldots,n} \{i/n - p_{(i)}\}, \quad D^- = \max_{i=1,\ldots,n} \{p_{(i)} - (i-1)/n\},$$

and reject normality if D is too large, see Fig. 4.2 for an illustration.

The V test from Kuiper [38] also uses the two *supremum statistics* D^- and D^+ by considering $V = D^+ + D^-$ as test statistic, see e.g. [86, Sect. 5.1.2].

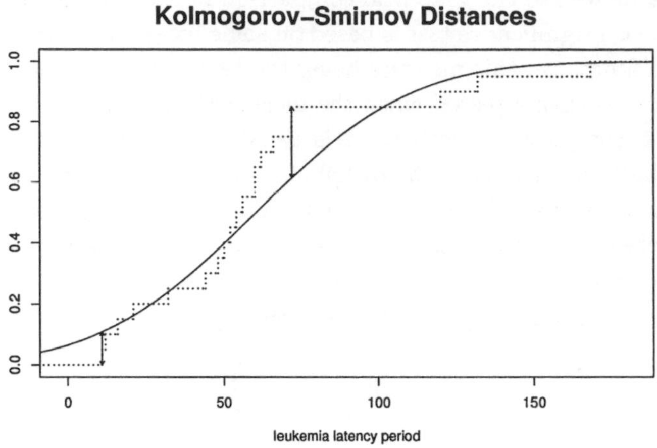

Kolmogorov–Smirnov Distances

Fig. 4.2: Differences D^- and D^+ for the leukemia latency period data from Table 3.3

P-values

Lilliefors [43] was presumably the first who noted that the distribution of D is different when μ and σ^2 are replaced by estimators, so that critical values indicating what 'too large' actually means change, compared to the situation when μ and σ^2 are assumed to be known. Lilliefors obtained critical values on the basis of simulation.

Dallal and Wilkinson [19] give a formula for the p-value of the Lilliefors test, which, however is claimed to be only reliable when the p-value is smaller than 0.1. In this case, if $5 \leq n \leq 100$, then the p-value p_* can be computed from

$$
\begin{aligned}
p_* = \exp(&-7.01256D^2(n + 2.78019) \\
&+ 2.99587D(n + 2.78019)^{1/2} - 0.122119 \\
&+ 0.974598/n^{1/2} + 1.67997/n) \, .
\end{aligned}
$$

For a sample size $n > 100$ replace D by $D(n/100)^{49}$ and then apply the above formula by replacing n by the number 100.

Alternatively, one may compare the value of the modified test statistic

$$D^* = D(\sqrt{n} - 0.01 + 0.85/\sqrt{n})$$

with the critical value according to our Table 4.2 taken from Stephens [81, Table 4.7], and reject the hypothesis of normality at level α when D^* is greater than the corresponding critical value $c(\alpha)$.

	significance level α				
	0.15	0.1	0.05	0.025	0.01
critical value $c(\alpha)$	0.775	0.819	0.895	0.995	1.035

Table 4.2: Critical values for D^* from Stephens [81, Table 4.7]

Example 4.2 Consider again the $n = 100$ female athletes weights from Table 1.1. The values of D^+ and D^- are almost identical here, and $D = 0.0547$. The computed p-value from the Dallal-Wilkinson approximation is $p_* = 0.6054$, being not reliable, but indicating that the hypothesis of normality is not rejected at level $\alpha = 0.1$ (or smaller). In addition, the modified statistic reads $D^* = 0.5509$, and Table 4.2 confirms that the hypothesis cannot be rejected at level $\alpha = 0.15$ (or smaller), since $D^* < 0.775$.   □

Recommendations

The Lilliefors (Kolmogorov-Smirnov) test is usually *not* recommended for testing normality. D'Agostino [1, p. 406] emphasizes that:

> For testing for normality, the Kolmogorov-Smirnov test is only a historical curiosity. It should never be used.

Stephens [81, p. 167] notes:

> The most famous statistic, the Kolmogorov-Smirnov D, tends to be weak in power.

4.3.2   The Cramér-von Mises Test

Anderson and Darling [4] proposed tests based on the measure

$$n \int_{-\infty}^{\infty} [F_n(x) - F(x)]^2 \psi(F(x)) \, dF(x) \, ,$$

where $\psi(F(x))$ is a weight function. Tests using such a measure are also called *quadratic ecdf* tests, since they involve the squared difference $[F_n(x) - F(x)]^2$. By choosing $F(x)$ as the cumulative distribution function of $N(\overline{x}, s^2)$ and $\psi(F(x)) = 1$, the above measure becomes

$$W^2 = \frac{1}{12n} + \sum_{i=1}^{n} \left(p_{(i)} - \frac{2i-1}{2n} \right)^2 \, .$$

This is the so-called Cramér-von Mises test statistic. The hypothesis of normality is rejected, when W^2 is too large. Fig. 4.3 illustrates the Cramér-von Mises distances $p_{(i)} - (2i-1)/2n$ for the leukemia latency period data.

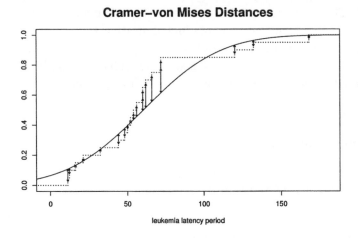

Fig. 4.3: Individual Differences $|p_{(i)} - (2i-1)/2n|$ for the leukemia latency period data from Table 3.3

P-values

Stephens [81, Table 4.9] gives approximations for p-values being valid for all sample sizes n. These formulas are given in Table 4.3.

$z = W^2(1.0 + 0.5/n)$	p-value
$z \leq 0.0275$	$1 - \exp(-13.953 + 775.5z - 12542.61z^2)$
$0.0275 < z \leq 0.051$	$1 - \exp(-5.903 + 179.546z - 1515.29z^2)$
$0.051 < z \leq 0.092$	$\exp(0.886 - 31.62z + 10.897z^2)$
$0.092 < z$	$\exp(1.111 - 34.242z + 12.832z^2)$

Table 4.3: Formulas for p-values of the Cramér-von Mises test for normality

Example 4.3 For the $n = 100$ female athletes weights, the Cramér-von Mises test statistic computes to $W^2 = 0.06$. Then the modified statistic $z = W^2(1.0+0.5/n)$ is $z = 0.06033$, and according to Table 4.3, the p-value is computed as $p_* = \exp(0.886 - 31.62z + 10.897z^2) = 0.3746$. □

4.3.3 The Anderson-Darling Test

When $F(x)$ is the cumulative distribution function of $N(\bar{x}, s^2)$ and

$$\psi(F(x)) = [F(x)(1 - F(x))]^{-1} ,$$

the above quadratic ecdf test statistic becomes

$$A^2 = -n - \frac{1}{n} \sum_{i=1}^{n} [2i - 1][\ln(p_{(i)}) + \ln(1 - p_{(n-i+1)})] ,$$

see [81, Sect. 4.2.2] for details. The hypothesis of normality is rejected when A^2 is too large.

P-values

Approximations for p-values can be obtained from Table 4.9 in [81]. See our Table 4.4.

$z = A^2(1.0 + 0.75/n + 2.25/n^2)$	p-value
$z \leq 0.2$	$1 - \exp(-13.436 + 101.14z - 223.73z^2)$
$0.2 < z \leq 0.34$	$1 - \exp(-8.318 + 42796z - 59.938z^2)$
$0.34 < z \leq 0.6$	$\exp(0.9177 - 4.279z - 1.38z^2)$
$0.6 < z$	$\exp(1.2937 - 5.709z + 0.0186z^2)$

Table 4.4: Formulas for p-values of the Anderson-Darling test for normality

Example 4.4 For the $n = 100$ female athletes weights, the Anderson-Darling test statistic computes to $A^2 = 0.3978$. Then the modified statistic $z = A^2(1.0 + 0.75/n + 2.25/n^2)$ is $z = 0.4009$, and according to Table 4.4, the p-value is computed as $p_* = \exp(0.9177 - 4.279z - 1.38z^2) = 0.3608.\square$

Recommendation

Stephens [81, p. 167] recommends the Anderson-Darling test for testing for normality:

> From the power studies for tests for normality and exponentiality it appears that A^2 (or W^2 as second choice) should be the recommended omnibus test statistic for EDF tests with unknown parameters, with good power against a wide range of alternatives.

4.4   Correlation and Regression Tests

A normal quantile-quantile plot is a graphical aid for the user to judge the strongness of the linear relationship between the sample ordered values and the expected values of standard normal order statistics. The stronger the relationship, the more convincing is the assumption of normality. Tests based on correlation and regression provide measures and corresponding critical values for assessing the degree of linearity.

4.4.1 The Shapiro-Francia Test

The Shapiro-Francia [76] test is strongly related to the graphical procedure in Sect. 3.2.3, being based on the assumed linear relationship between the ordered observations $x_{(i)}$ and the expectations $m_i = \mathrm{E}(Z_{(i)})$, $i = 1, \ldots, n$, where $Z_1, \ldots, Z_n$ is a sample from the $N(0,1)$ distribution.

Instead of inspecting the degree of linearity by plotting the individual $x_{(i)}$ and m_i against each other, the (empirical) Pearson correlation $r(\vec{x}, \vec{m})$ between the vectors $\vec{x} = (x_{(1)}, \ldots, x_{(n)})$ and $\vec{m} = (m_1, \ldots, m_n)$ can be computed as

$$r(\vec{x}, \vec{m}) = \frac{\sum_{i=1}^{n}(x_{(i)} - \overline{x})(m_i - \overline{m})}{\sqrt{\sum_{i=1}^{n}(x_{(i)} - \overline{x})^2}\sqrt{\sum_{i=1}^{n}(m_i - \overline{m})^2}} .$$

If the given data $x_1, \ldots, x_n$ does in fact come from a $N(\mu, \sigma^2)$ distribution, $r(\vec{x}, \vec{m})$ should be quite high, while otherwise it should be low. The Shapiro-Francia test statistic

$$W' = r(\vec{x}, \vec{m})^2$$

is simply the square of the Pearson correlation between $\vec{x}$ and $\vec{m}$, and is thus easier to compute than the statistic of the Shapiro-Wilk test [75] discussed in the following section, see also [86, Sect. 2.3]. Usually for computing W', the approximation

$$m_i = \Phi^{-1}\left(\frac{i - 3/8}{n + 1/4}\right)$$

is chosen.

The statistic W' can only take values between 0 and 1, and the hypothesis of normality is rejected when W' is too small. This test may equivalently be transformed into a right-tailed level-α-test when the statistic

$$Z' = n(1 - W')$$

is considered, and the hypothesis of normality is rejected when Z' is too large, see also [82].

P-values

Now, what does 'too small' for W', or equivalently 'too large' for Z' actually mean? Royston [72] gives a simple formula for obtaining p-values for the Shapiro-Francia test. These may be computed from the transformed statistic

$$U' = \left(\ln(1 - W') - \widehat{\alpha}\right) / \widehat{\beta} \, ,$$

being approximately $N(0,1)$ distributed. From [72],

$$\widehat{\alpha} = -1.2725 + 1.0521(v - u) \quad \text{and} \quad \widehat{\beta} = 1.0308 - 0.26758(v + 2/u) \, ,$$

where $u = \ln(n)$ and $v = \ln(u)$.

Example 4.5 For the $n = 100$ female athletes weights, the Pearson correlation $r(\vec{x}, \vec{m})$ computes to $r(\vec{x}, \vec{m}) = 0.9942$. Then $W' = 0.9885$, $Z' = 1.1548$, and $U' = 0.0980$. The p-value can be computed from U' as $p_* = 1 - \Phi(0.0980) = 0.461$.   □

Recommendation

Using the above formula for the computation of p-values, the Shapiro-Francia test has the advantage of being easy to implement. Moreover, Royston [72] comments:

> The statistic W' is one of the most powerful 'omnibus' (all-purpose) tests of non-normality available, so the test above is quick but not inferior.

4.4.2   The Shapiro-Wilk Test

A more known test procedure being related to the graphical procedure in Sect. 3.2.3, is the so-called *Shapiro-Wilk test* (also called *Shapiro test*, *W test*, or *analysis of variance test*). As the Shapiro-Francia test, it is based on the assumed linear relationship between the ordered observations $x_{(i)}$ and the expectations $m_i = \mathrm{E}(Z_{(i)})$, $i = 1, \ldots, n$, where $Z_1, \ldots, Z_n$ is a sample from the $N(0,1)$ distribution. The actual test statistic is derived via a regression model

$$X_{(i)} = \mu + \sigma m_i + \varepsilon_i, \quad i = 1, \ldots, n \, ,$$

where $X_{(i)}$ is the random variable admitting observation $x_{(i)}$, and ε_i is a random error. This model, however, is not a classical regression model since the errors ε_i are not uncorrelated. To the contrary,

$$\mathrm{Cov}(\varepsilon_i, \varepsilon_j) = \sigma^2 v_{ij}, \quad v_{ij} = \mathrm{Cov}(Z_{(i)}, Z_{(j)}) ,$$

where v_{ij} is not necessarily equal to zero for $i \neq j$. (In addition, v_{ii} and v_{kk} are not necessarily identical for $i \neq k$.) Based on generalized (Aitken) least-squares estimates for μ and σ, the value of the Shapiro-Wilk test statistic, judging the adequacy of the linear fit, is

$$W = \frac{\left(\sum_{i=1}^{n} a_i x_{(i)} \right)^2}{\sum_{i=1}^{n} (x_i - \overline{x})^2} ,$$

where each a_i depends on the m_i and v_{ij}, $i, j = 1, \ldots, n$. See [75, Sect. 2.2] or [86, Sect. 2.3] for details.

Unfortunately, the elements v_{ij} are only known for samples from size 3 up to size 20. For sample sizes up to 50, estimates of the a_i are given in [75]. Fortunately, Royston [71] extends the application of W to sample sizes between 7 and 2000.

Recommendation

The Shapiro-Wilk test is known to have good power properties against a wide range of alternatives. It is implemented in many statistical software packages and thus widely applied. On the other hand, the test statistic is more difficult to elucidate (and more difficult to implement) than the Shapiro-Francia test statistic, having comparable properties.

4.5   Summary

Tests for normality can be seen as data analytic tools for checking for normality without laying the claim to them as 'statistical proofs'. Then it is advisable to use phrases like 'the computed test statistic does not support the hypothesis of normality' or 'the computed test statistic gives no evidence for non-normality' rather than to speak of 'significant' or 'non-significant' test results. Since a test statistic reduces the complete data set to a single number, graphical analysis of all observations should complement the procedure.

As noted above, the two most famous omnibus tests (Lilliefors (more known as Kolmogorov-Smirnov) test and chi-square test) are *not* advisable for testing for normality. Instead, one may apply the Cramér-von Mises test, the Anderson-Darling test, the Shapiro-Francia test, being easy to implement with the above formulas for computing p-values, or the Shapiro-Wilk test, being already implemented in many statistical software packages.

Chapter 5

Variants of the Normal Distribution

When observations $x_1, \ldots, x_n$ of a specific variable X are given, then for different reasons there could be evidence that the distribution of X is not normal, but nonetheless related to the normal distribution. Three instances of non-normal distributions (the truncated normal, the two-normals-mixture and the skew-normal distribution), each being related to the normal distribution in its own specific way, are discussed in the subsequent Sections 5.1, 5.2 and 5.3.

5.1 Truncated Normal Distribution

Suppose that a random variable X is known to follow a distribution with pdf $f(x)$, but with the additional restriction that X can only take values in a given interval $[a, b]$. If $f(x)$ does not allow for the latter restriction, then, in view of the additional information, it cannot be the correct pdf of X. Instead, one may consider the conditional pdf

$$f(x|x \in [a, b]) \, ,$$

being the probability density function $f(x)$ of X given $X \in [a, b]$. An intuitive approach to obtain such a conditional pdf is via conditional probabilities. According to Remark 1.3

$$f(x)\,\Delta x \approx \mathrm{P}\left(X \in [x - \tfrac{1}{2}\,\Delta x, x + \tfrac{1}{2}\,\Delta x]\right),$$

for a small Δx. Then the conditional probability[1]

$$\mathrm{P}\left(X \in [x - \tfrac{1}{2}\,\Delta x, x + \tfrac{1}{2}\,\Delta x]\right)$$

given $X \in [a, b]$ is

$$\frac{\mathrm{P}\left(X \in [x - \tfrac{1}{2}\,\Delta x, x + \tfrac{1}{2}\,\Delta x] \quad \text{and} \quad X \in [a, b]\right)}{\mathrm{P}\left(X \in [a, b]\right)}.$$

The denominator can be expressed as $F(b) - F(a)$, where $F(\cdot)$ is the cdf corresponding to $f(\cdot)$, while the nominator is approximately $f(x)\,\Delta x\, I_{[a,b]}(x)$. Hence, naturally

$$f(x|x \in [a, b]) = \frac{f(x) I_{[a,b]}(x)}{F(b) - F(a)}.$$

The right-hand side is the general form of a pdf truncated on the left at a and on the right at b. It is simply the original pdf considered on the interval $[a, b]$ and rescaled by $1/(F(b) - F(a))$, in order to guarantee an overall probability mass 1 on $[a, b]$.

Such a truncated distribution is of practical relevance when it is known that a given random variable X is associated with a specific law of distribution $f(x)$, but for some reason, outcomes outside an interval $[a, b]$ cannot occur. One may also consider the truncation only on one side by either putting $a = -\infty$ or $b = \infty$.

5.1.1   Truncation of the Normal Distribution

Strictly speaking, most (if not all) random variables being assumed to follow a normal distribution, more realistically follow a *truncated normal* distribution. For example, the weight of a person cannot assume negative values with positive probability, while any normally distributed random variable does. Hence, the weight of a person cannot follow a normal distribution, but only a truncated normal distribution. Quite often, however, the truncation points a and b are outside the 3σ interval, i.e. $[\mu - 3\sigma, \mu + 3\sigma] \subset [a, b]$. In that case, if the random variable X is normally distributed, then it is very unlikely that X takes values outside $[\mu - 3\sigma, \mu + 3\sigma]$.

---

[1] We simply apply the rule $\mathrm{P}(A|B) = \mathrm{P}(A \cap B)/\mathrm{P}(B)$ for two events A and B.

Example 5.1 Consider the sum Y of $n = 10$ independent on $(0,1)$ uniformly distributed random variables, cf. Example 2.1. Then, Y takes its values in the interval $[0, n]$ with probability one. When approximating the distribution of Y by the $N(n/2, n/12)$ distribution, the corresponding 3σ interval is

$$[n/2 - 3\sqrt{n/12}, n/2 + 3\sqrt{n/12}] ,$$

being strictly contained in $[0, n]$ for $n > 3$. Hence, although values of Y outside $[0, n]$ cannot occur, the normal approximation for the distribution of Y is not invalidated by this fact. □

In such cases, the application of the truncated normal instead of the normal distribution would be an unnecessary complication of the situation from which no relevant gain in knowledge can be expected. The situation is different, however, when one of the truncation points is within the 3σ interval. For example, when we assume that the weight of a person is normally distributed with expectation $\mu = 74$ kilograms (say), but for some reason, our population consists only of persons with weight greater than 80 kilograms. Then it can be beneficial to consider the truncated normal distribution on $[80, \infty)$ with parameter $\mu = 74$ as a model for the distribution[2] in our population. Our new model assigns other probabilities to events in connection with the random variable than the old one. This is reflected by the fact that the new pdf now has an overall probability mass of 1 on the considered interval, while the normal distribution has an overall probability mass of 1 on the whole real line.

Probability Density Function

When X has an $N(\mu, \sigma^2)$ distribution, its pdf can be written as $f(x) = (1/\sigma)\phi[(x - \mu)/\sigma]$. Hence, in view of our previous considerations, the pdf of a truncated normal distribution is given as

$$f(x; \mu, \sigma) = \frac{1}{\sigma} \frac{\phi\left(\frac{x-\mu}{\sigma}\right)}{\left[\Phi\left(\frac{b-\mu}{\sigma}\right) - \Phi\left(\frac{a-\mu}{\sigma}\right)\right]} I_{[a,b]}(x)$$

$$= \frac{1}{\int_a^b e^{-\frac{1}{2}(t-\mu)^2/\sigma^2} \, dt} e^{-\frac{1}{2}(x-\mu)^2/\sigma^2} I_{[a,b]}(x) .$$

---

[2]Of course there is also a natural upper bound for the weight, but this may still be regarded as being too far away from μ to be of relevance.

In applications, the numbers a and b are usually known and thus not parameters. When $a = -\infty$, then the distribution is singly truncated from above, and when $b = \infty$ it is singly truncated from below. For the choices

$$a = \mu \quad \text{and} \quad b = \infty$$

it follows $\int_a^b e^{-\frac{1}{2}(t-\mu)^2/\sigma^2}\,dt = \frac{1}{2}\sqrt{2\pi}\sigma$, so that

$$f(x; \mu, \sigma) = \frac{2}{\sqrt{2\pi}\sigma}\, e^{-\frac{1}{2}(x-\mu)^2/\sigma^2} I_{[\mu,\infty)}(x)\,,$$

being the pdf of a half-normal distribution with location parameter μ and scale parameter σ, see Sect. 8.2.1. Figure 5.1 shows different truncated pdfs obtained from the standard normal distribution.

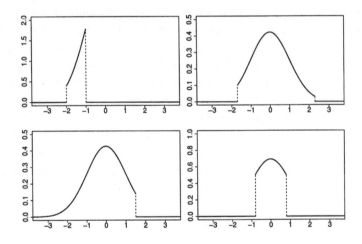

Fig. 5.1: Probability density functions of different truncations of the standard normal distribution

Expectation and Variance

Let us now assume that X follows a truncated normal distribution. Then expectation and variance of X are readily obtained, using e.g. some normal integral formulas. By applying the indefinite normal integral

$$\int x\,\phi(c+dx)\,\mathrm{d}x = -\frac{1}{d^2}\phi(c+dx) - \frac{c}{d^2}\Phi(c+dx)\,,$$

straightforward calculations yield

$$\mu_t := \mathrm{E}(X) = \mu - \sigma\frac{\phi\left(\frac{b-\mu}{\sigma}\right) - \phi\left(\frac{a-\mu}{\sigma}\right)}{\Phi\left(\frac{b-\mu}{\sigma}\right) - \Phi\left(\frac{a-\mu}{\sigma}\right)}\,.$$

The notation μ_t is chosen in order to distinguish the expectation of the truncated normal from the expectation μ of the non-truncated normal distribution. By applying the indefinite normal integral

$$\int x^2\phi(c+dx)\,\mathrm{d}x = \frac{c^2+1}{d^3}\Phi(c+dx) - \frac{dx-c}{d^3}\phi(c+dx)\,,$$

it follows

$$\mathrm{E}(X^2) = \mu^2 + \sigma^2 - \sigma\frac{(b+\mu)\phi\left(\frac{b-\mu}{\sigma}\right) - (a+\mu)\phi\left(\frac{a-\mu}{\sigma}\right)}{\Phi\left(\frac{b-\mu}{\sigma}\right) - \Phi\left(\frac{a-\mu}{\sigma}\right)}\,.$$

By letting $u = (a-\mu)/\sigma$ and $v = (b-\mu)/\sigma$, we may also write

$$\mathrm{E}(X^2) = \sigma^2 - \sigma^2\frac{v\phi(v) - u\phi(u)}{\Phi(v) - \Phi(u)} + 2\mu\mu_t - \mu^2\,.$$

From this, the variance of X can be obtained via $\mathrm{Var}(X) = \mathrm{E}(X^2) - [\mathrm{E}(X)]^2$ as

$$\sigma_t^2 := \mathrm{Var}(X) = \sigma^2\left[1 - \frac{v\phi(v) - u\phi(u)}{\Phi(v) - \Phi(u)}\right] - (\mu_t - \mu)^2\,.$$

5.1.2 Estimation of Parameters

To find estimates for μ and σ, the left-hand sides in the formulas for μ_t and σ_t^2 are replaced by $\overline{x} = \frac{1}{n}\sum_{i=1}^n x_i$ and $d^2 = \frac{1}{n}\sum_{i=1}^n (x_i - \overline{x})^2$, and then the two equations are solved for μ and σ^2. These solutions, being functions of $\overline{x}$ and d^2, are the *method of moments* estimates for μ and σ^2. They are known to coincide with the respective maximum likelihood estimates, see e.g. [73, Chapter 3].

There are two problems with the maximum likelihood estimates. First, they cannot be given in explicit form, since the equations

$$\bar{x} = \mu_t \quad \text{and} \quad d^2 = \sigma_t^2$$

are not linear in μ and σ^2. Hence, one may obtain estimates only by solving the equations numerically for a given sample $x_1, \ldots, x_n$.

Second, as pointed out in [50, 32], the estimates are not necessarily finite, meaning that the likelihood function can possibly become maximal for $(\mu, \sigma) = (\infty, \infty)$ or $(\mu, \sigma) = (-\infty, \infty)$. In that case the estimates are called non-existent. As a sufficient condition for the non-existence of the maximum likelihood estimates in the doubly truncated case (a and b both finite), Mittal and Dahiya [50] obtain

$$\frac{(b-a)^2}{12} < d^2 \ .$$

Hegde and Dahiya [32, Corollary 1] give a necessary and sufficient condition for the existence of the maximum likelihood estimates.

To cover the case of non-existence, Mittal and Dahiya [50] propose to solve the modified equations

$$\bar{x} = \mu_t \quad \text{and} \quad d^2 + \frac{1}{n} = \frac{2}{n}\sigma^2 + \sigma_t^2 \ .$$

numerically with respect to μ and σ^2.

The problem of non-existence of the maximum likelihood estimates is less pronounced for large sample sizes n and small degrees of truncation.

The EM Algorithm

As a rather simple way to obtain estimates, however not covering the possibility of non-existence, we consider the maximum likelihood estimation via the EM (expectation maximization) algorithm, see e.g. [20, 34, 21]. This is given in [73, p. 37/38], apart from some errors being corrected in the following description.

Given data points $x_1, \ldots, x_n$ from a truncated normal distribution with left truncation point a and right truncation point b, the initial estimates for μ and σ are

$$\mu^{(0)} = \bar{x} \quad \text{and} \quad \sigma^{(0)} = (\overline{x^2} - \bar{x}^2)^{1/2} \ ,$$

where $\overline{x} = \frac{1}{n}\sum_{i=1}^{n} x_i$ and $\overline{x^2} = \frac{1}{n}\sum_{i=1}^{n} x_i^2$. Let $\mu^{(k)}$ and $\sigma^{(k)}$ be the estimates of μ and σ at the k-th iteration. Compute

$$u^{(k)} = \frac{a - \mu^{(k)}}{\sigma^{(k)}}, \qquad v^{(k)} = \frac{b - \mu^{(k)}}{\sigma^{(k)}},$$

$$l^{(k)} = \text{int}\left[\frac{n\,\Phi(u^{(k)})}{\Phi(v^{(k)}) - \Phi(u^{(k)})}\right], \qquad r^{(k)} = \text{int}\left[\frac{n\,(1 - \Phi(v^{(k)}))}{\Phi(v^{(k)}) - \Phi(u^{(k)})}\right],$$

where $\text{int}[\cdot]$ is the integer part of a real number. Further, compute

$$\mu_a^{(k)} = \mu^{(k)} - \sigma^{(k)}\frac{\phi(u^{(k)})}{\Phi(u^{(k)})}, \qquad \mu_b^{(k)} = \mu^{(k)} + \sigma^{(k)}\frac{\phi(v^{(k)})}{1 - \Phi(v^{(k)})},$$

as well as

$$\mu_a^{2\,(k)} = \sigma^{(k)}[\sigma^{(k)} + u^{(k)}(\mu_a^{(k)} - \mu^{(k)})] + \mu^{(k)}[2\mu_a^{(k)} - \mu^{(k)}],$$

and

$$\mu_b^{2\,(k)} = \sigma^{(k)}[\sigma^{(k)} + v^{(k)}(\mu_b^{(k)} - \mu^{(k)})] + \mu^{(k)}[2\mu_b^{(k)} - \mu^{(k)}].$$

By letting $s^{(k)} = n + l^{(k)} + r^{(k)}$, the estimates of the $k + 1$ iteration are

$$\mu^{(k+1)} = \frac{n}{s^{(k)}}\overline{x} + \frac{l^{(k)}}{s^{(k)}}\mu_a^{(k)} + \frac{r^{(k)}}{s^{(k)}}\mu_b^{(k)},$$

and

$$\sigma^{(k+1)} = \left(\frac{n}{s^{(k)}}\overline{x^2} + \frac{l^{(k)}}{s^{(k)}}\mu_a^{2\,(k)} + \frac{r^{(k)}}{s^{(k)}}\mu_b^{2\,(k)} - (\mu^{(k+1)})^2\right)^{1/2}.$$

The iteration process can be continued until the absolute differences between the estimates of the k-th and $(k+1)$-th iteration are all smaller than 0.00001 (say) or the number of iterations exceeds 1000 (say).

The EM algorithm can also be applied to estimate the parameters of a singly truncated normal distribution. In that case the user should set a or b to some value being outstanding enough to model $a = -\infty$ or $b = \infty$.

Example 5.2 Figure 5.2 shows the histograms of four samples of size $n = 100$ from a truncated normal distribution with $a = 2$ and $b = 5$. The true parameters $\mu = 3.5$ and $\sigma^2 = 1$ are reflected by the true pdf as a dotted line. The estimated pdf obtained from the EM algorithm is given as a solid line. It can be seen that there are some differences between estimated and true pdf, but the fit to the data seems to be adequate in all four samples. A reason for this can be that, due to randomness, the sample does not always sufficiently reflect the true distribution. See also our discussion in the paragraph 'Preciseness of Estimation and Goodness of Fit', p. 41.   □

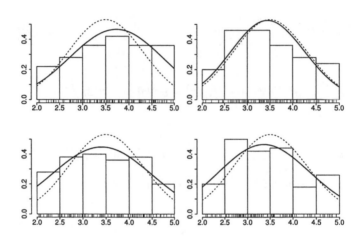

Fig. 5.2: Histograms of four samples of size n with estimated pdf (solid line) and true pdf (dotted line)

5.1.3   Random Sample Generation

To generate a pseudo-random sample from the truncated normal distribution, the classical method based on the *probability integral transformation* theorem, see [54, Sect. V 5.2], can be used. This method allows the generation of a sample from a continuous cumulative distribution function $F(x)$ when the generation of a sample from the uniform distribution is available, see also e.g. [70, Sect. 5.1].

Probability Integral Transformation

From the probability integral transformation theorem, if U is uniformly distributed over the interval $(0, 1)$, i.e. U has pdf $f(u) = I_{(0,1)}(u)$, the random variable $X = F^{-1}(U)$ has cumulative distribution function $F(\cdot)$. Hence, if $u_1, \ldots, u_n$ is a sample value from the uniform distribution on $(0, 1)$,

$$F^{-1}(u_1), \ldots, F^{-1}(u_n)$$

is a sample value from the distribution having cdf $F(\cdot)$.

Truncated Normal Random Sample

For the truncated normal distribution, the cumulative distribution function is given by

$$F(x) = \begin{cases} 0 & x < a \\ \dfrac{\Phi\left(\frac{x-\mu}{\sigma}\right) - \Phi\left(\frac{a-\mu}{\sigma}\right)}{\Phi\left(\frac{b-\mu}{\sigma}\right) - \Phi\left(\frac{a-\mu}{\sigma}\right)} & a \le x \le b \\ 1 & x > b \end{cases}.$$

Suppose now that u is an observation from the uniform distribution on $(0, 1)$. Then $x = F^{-1}(u)$ is the solution to the equation

$$\Phi\left(\frac{x - \mu}{\sigma}\right) = u\left[\Phi\left(\frac{b - \mu}{\sigma}\right) - \Phi\left(\frac{a - \mu}{\sigma}\right)\right] + \Phi\left(\frac{a - \mu}{\sigma}\right)$$

with respect to x. It may be obtained by the quantile function of the normal distribution. Generating $u_1, \ldots, u_n$ independently and solving the respective equations yields a sample $x_1, \ldots, x_n$ from the truncated normal distribution with given a, b as well as μ and σ.

5.2 Two-Normals-Mixture Distribution

When we have k probability density functions $f_i(x)$, $i = 1, \ldots, k$, then the combination

$$f(x) = \sum_{i=1}^{k} \alpha_i f_i(x)$$

with $\alpha_i \geq 0$ and $\sum_{i=1}^{k} \alpha_i = 1$ is again a probability density function. It is called a *mixture* of the individual densities. Mixtures can be considered for a variety of different distributions involved.

5.2.1   Probability Density Function

The mixture of $k = 2$ normal densities, has pdf

$$f(x; \alpha, \mu_1, \mu_2, \sigma_1, \sigma_2) = \alpha f_1(x; \mu_1, \sigma_1) + (1 - \alpha) f_2(x; \mu_2, \sigma_2) ,$$

where

$$f_j(x; \mu_j, \sigma_j) = \frac{1}{\sigma_j} \phi\left(\frac{x - \mu_j}{\sigma_j}\right) , \quad j = 1, 2 .$$

The corresponding distribution is also called a *contaminated normal distribution* when α is close to 1. In that case, the distribution can be taken as a model for data containing outliers, see also Example 5.3. We will, however, not assume that α is necessarily close to 1 and therefore we will call it the *two-normals-mixture distribution*.

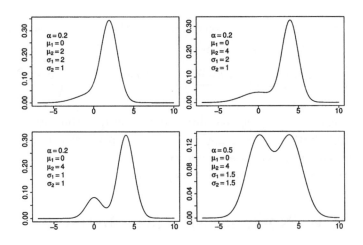

Fig. 5.3: Probability density functions of the two-normals-mixture distribution for different parameter values

The pdf of a two-normals-mixture distribution can have one but also two modes, depending on the parameters, see Fig. 5.3. In case

$$\mu_1 = \mu_2 =: \mu\,,$$

the pdf is symmetric about μ (with mode μ). Therefore, the two-normals-mixture distribution may be used as a model for a distribution with a heavier tail, see also Example 5.3, as well as a model for bimodal data.

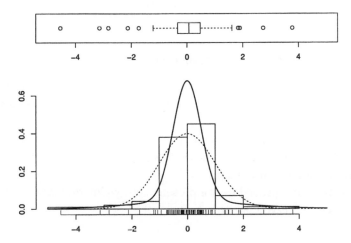

Fig. 5.4: Boxplot and histogram of a sample of size $n = 100$ from a two-normals-mixture distribution given in Example 5.3, superimposed by the corresponding pdf (solid line) and the pdf of the standard normal distribution (dotted line)

Expectation and Variance

When X has a two-normals-mixture distribution, then

$$E(X) = \alpha\mu_1 + (1 - \alpha)\mu_2$$

and

$$E(X^2) = \alpha(\mu_1^2 + \sigma_1^2) + (1 - \alpha)(\mu_2^2 + \sigma_2^2)\,,$$

from which one may also obtain a formula for $\text{Var}(X) = E(X^2) - [E(X)]^2$.

Example 5.3 Consider a two-normals-mixture distribution with parameters $\alpha = 0.8$, $\mu_1 = \mu_2 = 0$, $\sigma_1 = 0.5$, and $\sigma_2 = 2$. The distribution is symmetric about zero, having expectation 0 and variance 1. When a sample comes from this distribution, most values will be near zero, but some will be rather far away from the bulk of the data. See Fig. 5.4, showing a boxplot and a histogram of such a sample of size $n = 100$, the latter being superimposed by the pdf of the corresponding two-normals-mixture distribution (solid line) and by the pdf of the standard normal distribution (dotted line). □

5.2.2   Estimation of Parameters

When we have observations $x_1, \ldots, x_n$, being assumed to come from a two-normals-mixture distribution, the estimation of the five parameters α, μ_1, μ_2, σ_1, σ_2 is of interest.

Presumably the earliest approach can de dated back to Pearson [64], who considered estimation by the *methods of moments* principle. See also [36, Sect. 10.2] for a discussion.

The estimation procedure we present here yields maximum likelihood estimates via the EM algorithm.

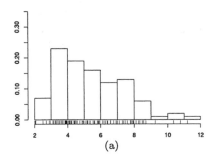 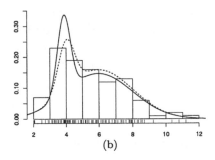

Fig. 5.5: Histogram of $n = 100$ data points (a) and superimposed by estimated (solid line) and true (dotted line) probability density function (b)

The EM Algorithm

The algorithm requires the specification of initial values $\alpha^{(0)}$, $\mu_1^{(0)}$, $\mu_2^{(0)}$, $\sigma_1^{(0)}$, $\sigma_2^{(0)}$ by the user. Starting from these, the estimates are iteratively computed as follows.

Let $\alpha^{(k)}$, $\mu_1^{(k)}$, $\mu_2^{(k)}$, $\sigma_1^{(k)}$, $\sigma_2^{(k)}$ be the estimates at the k-th iteration. Compute

$$f_j^{(k)}(x_i) := f_j(x_i; \mu_j^{(k)}, \sigma_j^{(k)}), \quad j = 1, 2,$$

$$f^{(k)}(x_i) := \alpha^{(k)} f_1^{(k)}(x_i) + (1 - \alpha^{(k)}) f_2^{(k)}(x_i),$$

and

$$w_{i1}^{(k)} = \alpha^{(k)} f_1^{(k)}(x_i)/f^{(k)}(x_i), \quad w_{i2}^{(k)} = (1 - \alpha^{(k)}) f_2^{(k)}(x_i)/f^{(k)}(x_i)$$

for $i = 1, \ldots, n$. Then the estimates at the $k + 1$ iteration are

$$\mu_1^{(k+1)} = \sum_{i=1}^{n} w_{i1}^{(k)} x_i \left/ \sum_{i=1}^{n} w_{i1}^{(k)} \right.,$$

$$\mu_2^{(k+1)} = \sum_{i=1}^{n} w_{i2}^{(k)} x_i \left/ \sum_{i=1}^{n} w_{i2}^{(k)} \right.,$$

$$\sigma_1^{(k+1)} = \sqrt{\sum_{i=1}^{n} w_{i1}^{(k)} (x_i - \mu_1^{(k)})^2 \left/ \sum_{i=1}^{n} w_{i1}^{(k)} \right.},$$

$$\sigma_2^{(k+1)} = \sqrt{\sum_{i=1}^{n} w_{i2}^{(k)} (x_i - \mu_2^{(k)})^2 \left/ \sum_{i=1}^{n} w_{i2}^{(k)} \right.},$$

and

$$\alpha^{(k+1)} = \frac{1}{n} \sum_{i=1}^{n} w_{i1}.$$

The iteration process can be continued until the absolute differences between the estimates of the k-th and $(k+1)$-th iteration are all smaller than 0.00001 (say) or the number of iterations exceeds 1000 (say).

The iteration is easy to program and works fast. The algorithm does, however, not in any case yield the proper estimates. The result may also depend on the choice of initial values. There are two strategies, usually giving reliable estimates:

- Use admissible and reasonable initial values. It is clear that $\alpha^{(0)}$ should be chosen between 0 and 1, but values too close to 0 or 1 should be avoided. The location parameters $\mu_1^{(0)}$ and $\mu_2^{(0)}$ should clearly be within the range of the observed data points. They should be neither too close nor too far away from each other. Initial guesses can in most cases be drawn from a histogram of the data. The parameters $\sigma_1^{(0)}$ and $\sigma_2^{(0)}$ should of course be greater than but also not too close to zero.

- Use different sets of starting values and see the effect on the estimates. Usually a good fit to the data (compared with a corresponding histogram) indicates that the obtained estimates are the desired ones.

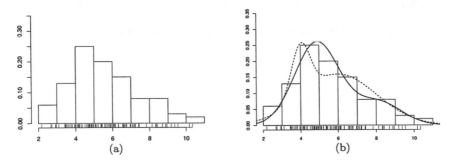

Fig. 5.6: Histogram of $n = 100$ data points (a) and superimposed by estimated (solid line) and true (dotted line) probability density function (b)

Example 5.4 Consider $n = 100$ data points with corresponding histogram from Fig. 5.5 (a). We assume that the data comes from a two-normals-mixture distribution and wish to estimate the unknown parameters. To apply the above algorithm we need starting values $\alpha^{(0)}$, $\mu_1^{(0)}$, $\mu_2^{(0)}$, $\sigma_1^{(0)}$, $\sigma_2^{(0)}$. From the histogram, a subjective guess for the location parameters could be $\mu_1^{(0)} = 3.5$ and $\mu_2^{(0)} = 7$. Since we have no idea about α, we choose $\alpha^{(0)} = 0.5$. Eventually, since the histogram has a short left tail and a long right tail, we may conclude that σ_1 should be smaller than σ_2, but since we do not have any rule of choice available, we simply put

$\sigma_1^{(0)} = \sigma_2^{(0)} = s = 1.9985$. Then the application of the above algorithm yields

$$\widehat{\alpha} = 0.2505, \; \widehat{\mu}_1 = 3.8606, \; \widehat{\mu}_2 = 5.9660, \; \widehat{\sigma}_1 = 0.3989, \; \widehat{\sigma}_2 = 2.0279$$

as our estimates. Since we have simulated the data, we know that the true values are

$$\alpha = 0.2, \; \mu_1 = 4, \; \mu_2 = 6, \; \sigma_1 = 0.5, \; \sigma_2 = 2 \; .$$

As it seems, the estimated parameters are not too far away from the true ones. The most striking difference seems to be the underestimation of the parameter σ_1, which causes the estimated pdf to have a higher peak than the true one actually has, see Fig. 5.5 (b).

Let us now consider a second sample of size $n = 100$ from the same distribution as above, with data points and corresponding histogram as shown in Fig. 5.6 (a). Again we apply the above algorithm to obtain the maximum likelihood estimates. We choose initial values $\alpha^{(0)} = 0.5$, $\mu_1^{(0)} = 4.5$, $\mu_2^{(0)} = 7$, $\sigma_1^{(0)} = \sigma_2^{(0)} = s = 1.8084$. Then the computed estimates are

$$\widehat{\alpha} = 0.7780, \; \widehat{\mu}_1 = 4.8777, \; \widehat{\mu}_2 = 8.0983, \; \widehat{\sigma}_1 = 1.1967, \; \widehat{\sigma}_2 = 1.2237 \; .$$

These estimates appear to be not very close to the true parameters. This is, however not due to a failure of the algorithm or inappropriate initial values. Other sets of initial values yield the same result. As a matter of fact, from Fig. 5.6 (b) the fit of the estimated pdf appears to be quite good, so that we can be quite sure that the estimates are the maximum likelihood ones. The reason for this is, that, due to randomness, the sample does not sufficiently reflect the true distribution, but looks more like to come from the estimated distribution. □

	exam point																		
1 – 20	21	2	45	7	0	0	20	25	0	25	18	58	0	84	63	0	14	11	18	71
21 – 40	16	38	0	21	20	0	33	25	0	30	19	64	0	3	59	0	61	19	59	21
41 – 53	62	22	29	45	75	11	8	65	56	57	10	44	24							

Table 5.1: Written statistics exam points

Example 5.5 Table 5.1 shows the points, $n = 53$ students received in a written statistics exam out of 145 gettable points.

To obtain a first visual impression of the data, a *stem-and-leaf plot* is generated, grouping the exam points with respect to the first digit (the *stem*), and arranging the second digits (the *leaves*) behind the corresponding first digit[3].

```
0 | 00000000002378
1 | 011468899
2 | 00111245559
3 | 038
4 | 455
5 | 67899
6 | 12345
7 | 15
8 | 4
```

Fig. 5.7: Stem-and-leaf plot of the data from Table 5.1

There are 10 zero point results. Suppose that we are interested in the distribution of the more than zero point results. Clearly, the corresponding random variable is discrete, but due to the variety of achievable (and achieved) points may nonetheless be approximated by a probability density function. Ignoring zero point results, the stem-and-leaf plot would indicate a bimodal distribution. Fitting a two-normals-mixture distribution via the EM algorithm gives the estimated probability density function

$$\widehat{f}(x) = \widehat{\alpha}\,\frac{1}{\widehat{\sigma}_1}\phi\left(\frac{x - \widehat{\mu}_1}{\widehat{\sigma}_1}\right) + (1 - \widehat{\alpha})\,\frac{1}{\widehat{\sigma}_2}\phi\left(\frac{x - \widehat{\mu}_2}{\widehat{\sigma}_2}\right)\,,$$

where

$$\widehat{\alpha} = 0.6122,\ \widehat{\mu}_1 = 18.49,\ \widehat{\mu}_2 = 59.44,\ \widehat{\sigma}_1 = 8.31,\ \widehat{\sigma}_2 = 11.36\,.$$

When it is desired to restrict the support of the estimated pdf $\widehat{f}(x)$ to the interval $[0, 145]$, one may consider the truncated pdf

---

[3]Here, the decimal point is 1 digit to the right of the |, which is of course not necessarily the case for other types of data.

$$\widetilde{f}(x) = \frac{1}{\widehat{F}(145) - \widehat{F}(0)} \widehat{f}(x) I_{[0,145]}(x) \,,$$

where

$$\widehat{F}(x) = \widehat{\alpha}\Phi\left(\frac{x - \widehat{\mu}_1}{\widehat{\sigma}_1}\right) + (1 - \widehat{\alpha})\Phi\left(\frac{x - \widehat{\mu}_2}{\widehat{\sigma}_2}\right) \,.$$

Since $1/(\widehat{F}(145) - \widehat{F}(0)) = 1.008062$, there is only a marginal difference between $\widehat{f}(x)$ and $\widetilde{f}(x)$. See Fig. 5.8 for a histogram of the greater than zero exam point results, together with the estimated pdf $\widetilde{f}(x)$. $\qquad\square$

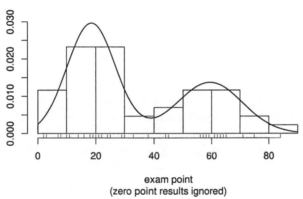

Fig. 5.8: Histogram and estimated probability density function of the data from Table 5.1

When it is desired to fit a *symmetric* two-normals-mixture distribution to given data, the above discussed EM algorithm may slightly be altered such that the parameters μ_1 and μ_2 are considered as only one parameter μ. Then a simple ad-hoc procedure is to estimate μ by the sample mean, not altering the estimate from iteration to iteration, while otherwise letting the algorithm unharmed.

Example 5.6 To determine the gas mileage ratings of a new car model, a sample of $n = 100$ cars is drawn with corresponding mileage observa-

	Miles per gallon									
1 – 10	36.3	32.7	40.5	36.2	38.5	36.3	41.0	37.0	37.1	39.9
11 – 20	41.0	37.3	36.5	37.9	39.0	36.8	31.8	37.2	40.3	36.9
21 – 30	36.9	41.2	37.6	36.0	35.5	32.5	37.3	40.7	36.7	32.9
31 – 40	37.1	36.6	33.9	37.9	34.8	36.4	33.1	37.4	37.0	33.8
41 – 50	44.9	32.9	40.2	35.9	38.6	40.5	37.0	37.1	33.9	39.8
51 – 60	36.8	36.5	36.4	38.2	39.4	36.6	37.6	37.8	40.1	34.0
61 – 70	30.0	33.2	37.7	38.3	35.3	36.1	37.0	35.9	38.0	36.8
71 – 80	37.2	37.4	37.7	35.7	34.4	38.2	38.7	35.6	35.2	35.0
81 – 90	42.1	37.5	40.0	35.6	38.8	38.4	39.0	36.7	34.8	38.1
91 – 100	36.7	33.6	34.2	35.1	39.7	39.3	35.8	34.5	39.5	36.9

Table 5.2: Gas mileage ratings for 100 cars, see [51, p. 235]

tions given in Table 5.2, see [51, p. 235]. Although the data admits some evidence that gas mileage rating can be assumed to be a normal variable, the histogram and the normal quantile-quantile plot indicate an unusual frequency peak around the center of the data, not very well in accordance with the normal distribution, see Fig. 5.9 (a) and (b). This causes us to try a symmetric two-normals-mixture distribution fit.

The mean of the data is $\bar{x} = 36.998$. The application of the above EM algorithm with initial values $\alpha^{(0)} = 0.5$, $\mu_1^{(0)} = \mu_2^{(0)} = \bar{x}$, $\sigma_1^{(0)} = 1$, and $\sigma_2^{(0)} = 2$, without altering the estimates for μ_1 and μ_2, yields

$$\hat{\alpha} = 0.1587, \ \hat{\mu}_1 = \hat{\mu}_2 = 36.994, \ \hat{\sigma}_1 = 0.3518, \ \hat{\sigma}_2 = 2.6185 \ .$$

The corresponding pdf is shown in Fig. 5.10. Since the (two parameter) normal distribution is a special case of the (four parameter) symmetric two-normals-mixture distribution, the actual AICs of the two fitted distributions may be compared with each other. The AIC for the former comes to 463.3673, while the AIC for the latter comes to 459.5441. Hence, the AIC criterion favors the fit by the two-normals-mixture distribution. On the other hand, the BIC for the normal fit is 468.5777, while the BIC for the two-normals-mixture fit is 469.9647, so that from the BIC the normal fit is preferred. □

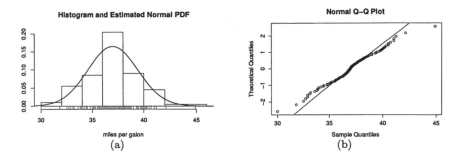

Fig. 5.9: Histogram with estimated normal pdf (a) and normal quantile-quantile plot (b) of data from Table 5.2

5.2.3 Random Sample Generation

To generate a pseudo-random sample from the two-normals-mixture distribution, one may use pseudo-random numbers from the binomial and the normal distribution, being usually available in a statistical software package. Given α, μ_1, σ_1, μ_2, σ_2, in a first step one may generate a value b, being identical to 1 with probability α and identical to 0 with probability $1 - \alpha$. (Then b is the outcome of a binomial distribution with parameters $n = 1$ and $p = \alpha$.) In a second step, if b turns out to be 1, then a random number x from the normal distribution with parameters μ_1 and σ_1^2 is generated, while otherwise a random number x from the normal distribution with parameters μ_2 and σ_2^2 is generated.

If these two steps are independently repeated n times, the obtained values $x_1, \ldots, x_n$ constitute a sample from the two-normals-mixture distribution.

5.3 Skew-Normal Distribution

The normal distribution with parameters μ and σ^2 is symmetric about μ. In order to adjust for a possible skewness in a given data set, it would be of interest to fit a more flexible pdf, coinciding with the normal pdf in case

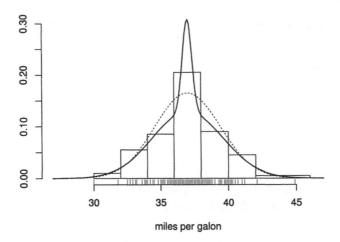

miles per galon

Fig. 5.10: Probability density function of the estimated symmetric two-normals-mixture distribution (solid line) together with estimated normal pdf (dotted line) for the data from Table 5.2

of a symmetric distribution of the data. Such a possibility is offered by the so-called *skew-normal distribution* introduced by Azzalini [6].

5.3.1   Probability Density Function

The random variable X is said to have a *skew-normal distribution* if its pdf is given by

$$f(x; \lambda) = 2\phi(x)\Phi(\lambda x), \quad -\infty < x < \infty,$$

for some scalar $-\infty < \lambda < \infty$, see [6]. By using the normal integral

$$\int_{-\infty}^{\infty} \Phi(a + bx)\phi(x)\, dx = \Phi(a/\sqrt{1 + b^2})$$

for $a = 0$ and $b = \lambda$, it follows that $\int_{-\infty}^{\infty} \phi(x)\Phi(\lambda x)\, dx = 1/2$ for any value of λ, showing that $f(x; \lambda)$ is in a fact a pdf[4]. For the choice $\lambda = 0$ the skew-normal distribution coincides with the standard normal distribution. For

---

[4]Azzalini [6] gives a different proof for a more general class of pdfs.

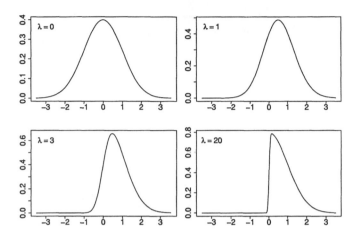

Fig. 5.11: Probability density function of the skew normal distribution for different values λ

$\lambda \to \infty$ the skew-normal distribution approaches the distribution of the absolute value of a random variable having standard normal distribution (half-normal distribution, see Sect. 8.2.1). Fig. 5.11 shows the pdf of the skew-normal distribution for different values of λ.

Adjusted Skew-Normal Distribution

When the skew-normal distribution is fitted to given data, it is necessary to adjust for location and scale. For any fixed value of λ the pdf $2\phi(x)\Phi(\lambda x)$ can be the standard pdf of a location-scale family of pdfs of the form

$$f(x; a, b, \lambda) = \frac{2}{b} \phi\left(\frac{x-a}{b}\right) \Phi\left(\lambda\frac{x-a}{b}\right) ,$$

where $-\infty < a < \infty$, $0 < b$, and $-\infty < \lambda < \infty$ are parameters. We call it the *adjusted skew-normal distribution*.

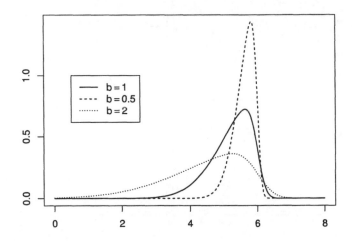

Fig. 5.12: Adjusted skew-normal distribution with parameters $a = 6$, $\lambda = -5$ and different choices for b

Expectation and Variance

In view of Theorem 2.2, a random variable X having adjusted skew-normal distribution with parameters a, b, and λ can be written as $X = a + bZ$, where Z has skew-normal distribution with parameter λ. It is known (see [6]) that

$$\mathrm{E}(Z) = \sqrt{\frac{2}{\pi}} \frac{\lambda}{\sqrt{1 + \lambda^2}} \quad \text{and} \quad \mathrm{Var}(Z) = 1 - [\mathrm{E}(Z)]^2 \;.$$

From this, the expectation $\mathrm{E}(X) = \mathrm{E}(a+bZ) = a+b\mathrm{E}(Z)$ and the variance $\mathrm{Var}(X) = \mathrm{Var}(a + bZ) = b^2\mathrm{Var}(Z)$ may easily be deduced.

Skewness

The coefficient of skewness of the adjusted skew-normal distribution is the same as that for the skew-normal distribution, given as

$$\sqrt{\beta}_1 = \frac{4 - \pi}{2} \frac{[\mathrm{E}(Z)]^3}{(1 - [\mathrm{E}(Z)]^2)^{3/2}} \;.$$

For $\lambda \to \infty$ it converges to

$$\sqrt{2}\frac{4-\pi}{(\pi-2)^{3/2}} \approx 0.995 \, ,$$

while for $\lambda \to -\infty$ it converges to the negative of this value.

5.3.2 Estimation of Parameters

To obtain maximum likelihood estimates of the parameters a, b and λ for a given sample value $x_1, \ldots, x_n$ from the adjusted skew-normal distribution, the log-likelihood

$$\ln\left[L(a, b, \lambda; x_1, \ldots, x_n)\right]$$
$$= -\ln(\sqrt{2\pi}) - n\ln(b) - \frac{1}{2b^2}\sum_{i=1}^{n}(x_i - a)^2 + n\ln(2) + \nu \, ,$$

where

$$\nu \equiv \nu(a, b, \lambda; x_1, \ldots, x_n) := \sum_{i=1}^{n}\ln\left[\Phi\left(\lambda\frac{x_i - a}{b}\right)\right] \, ,$$

may be maximized with respect to a, b and λ. For $\lambda = 0$, the last summand becomes $\nu = -n\ln(2)$, showing that in this case the log-likelihood coincides with the log-likelihood of the normal distribution with a and b replaced by μ and σ, respectively. In view of the presence of ν, the maximum likelihood estimates cannot be given in explicit form, but may only be obtained by some numerical optimization algorithm.

As two necessary conditions for a maximum in a, b (or b^2) and λ, the partial derivatives of the log-likelihood with respect to b^2 and λ must be equal to zero. Obviously, the partial derivative of the log-likelihood with respect to λ is the partial derivative $\partial\nu/\partial\lambda$. Since it can be shown that

$$\frac{\partial\nu}{\partial b^2} = -\frac{\lambda}{2b^2}\frac{\partial\nu}{\partial\lambda} \, ,$$

the partial derivative of the log-likelihood with respect to b^2 is

$$-\frac{n}{2b^2} + \frac{1}{2b^4}\sum_{i=1}^{n}(x_i - a)^2$$

when $\partial \nu / \partial \lambda = 0$. Putting this equal to zero shows that the maximum-likelihood estimates $\widehat{a}$ and $\widehat{b}$ for a and b must satisfy the relationship

$$\widehat{b} = \sqrt{\frac{1}{n} \sum_{i=1}^{n} (x_i - \widehat{a})^2}\,.$$

This may be imposed as an additional restriction in the numerical optimization algorithm, or it may be used to check the validity of an already obtained solution.

Real-Life Data and the Adjusted Skew-Normal Distribution

When data points are given and it comes to question whether the distribution of the underlying variable is symmetric or not, a real-life data set will almost always reveal at least a slight skewness. Hence, the problem of fitting either the normal or the adjusted skew-normal distribution can also be the question how to treat the trade-off between the fitting of a less flexible but possibly more meaningful pdf (the normal pdf), and the fitting of a more flexible but possibly less meaningful pdf (the adjusted skew-normal pdf). Strictly speaking, there may be instances when the fitting of the adjusted skew-normal distribution is an unnecessary complication for inferential purposes, while the fitting of the normal distribution is precise enough and easier to handle.

Example 5.7 For the $n = 100$ data points of female athletes weights from Table 1.1, the histogram (see also Fig. 1.4, p. 7) and the sample coefficient of skewness (see also Table 3.2, p. 91) indicate that the distribution is slightly skewed to the left. When we assume that the data comes from an adjusted skew-normal distribution, the maximum likelihood estimates of the parameters come up to

$$\widehat{a} = 75.96, \quad \widehat{b} = 13.86, \quad \widehat{\lambda} = -1.2426\,.$$

It is easily checked that the restriction $\widehat{b} = \sqrt{\frac{1}{n}\sum_{i=1}^{n}(x_i - \widehat{a})^2}$ is satisfied. Fig. 5.13 shows the corresponding fit and compares it with the fit obtained from the normal distribution. The difference between the two fits is only marginal. Since the (two-parameter) normal distribution is a special case of the (three parameter) adjusted skew-normal distribution, the actual

AICs of the fitted distributions may be compared with each other. The AIC of the normal distribution fit comes to 764.8235, while the AIC of the fitted adjusted skew-normal distribution comes to 766.1534, showing that the AIC criterion also favors the normal fit. □

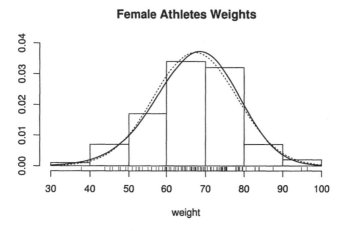

Fig. 5.13: Fit from the adjusted skew-normal distribution (solid line) and from the normal distribution (dotted line)

5.3.3 Random Sample Generation

To generate a pseudo-random sample from the adjusted skew-normal distribution, one may use pseudo-random numbers from the normal distribution. As a matter of fact, if U and V are two independent standard normal variables, then

$$Z = \frac{\lambda}{\sqrt{1 + \lambda^2}}|U| + \frac{1}{\sqrt{1 + \lambda^2}}V$$

has skew-normal distribution with parameter λ, see [33]. Hence, one may generate two independent pseudo-random numbers u and v from the standard normal distribution and compute $z = \frac{\lambda}{\sqrt{1+\lambda^2}}|u| + \frac{1}{\sqrt{1+\lambda^2}}v$. Then one may compute $x = a + bz$. Repeating this independently n times yields the desired sample $x_1, \ldots, x_n$.

Chapter 6

Transformations to Normality

When the distribution of a random variable X is not normal, then it might be possible to transform it to a variable whose distribution can in fact assumed to be normal.

6.1 Data Transformation

Suppose that we wish to investigate a random variable X from a corresponding random sample $X_1, \ldots, X_n$ with observations $x_1, \ldots, x_n$. The distribution of X is not normal, but otherwise not exactly known.

How can we handle such a situation? A simple approach is to assume that the random variable X, though not normal itself, possesses a 'normal kernel' in the sense that there is a function $h(\cdot)$ such that $h(X) = Z$ has a normal distribution. Then we may

- learn more about the function $h(\cdot)$ and use our knowledge to estimate the pdf of X. In Sections 6.2 and 6.3, specific pdfs of X being obtained from an assumed relationship $h(X) = Z$, where Z has standard normal distribution, are described. Alternatively, we may

- specify an appropriate function $h(\cdot)$ and carry out data analysis using the 'normal sample' $z_1, \ldots, z_n$, where $z_i = h(x_i)$. The drawn conclusions may then be retranslated to inference about the original variable X. In Sect. 6.4, so-called power transformations are discussed.

6.2   The Johnson System of Distributions

Given a random variable X, Johnson [35] considered a transformation $Z = h(X)$ having standard normal distribution, where $h(\cdot)$ is of the form

$$h(X) = \gamma + \delta g\left(\frac{X - \xi}{\lambda}\right) .$$

The quantities γ, δ, ξ, and λ are unknown parameters, where it is assumed that $\delta > 0$ and $\lambda > 0$. The function $g(\cdot)$ should be of a simple form, not depending on additional parameters, and must further be specified.

Under this assumption, it is of interest to learn about the distribution and pdf of the random variable X. By letting $Y = (X - \xi)/\lambda$, it follows that

$$Z = \gamma + \delta g(Y) ,$$

and for $g(\cdot)$ having a continuous derivative on the support[1] of Y, the pdf of Y is given as

$$f(y) = \delta \left|\frac{\mathrm{d}}{\mathrm{d}y}g(y)\right| \phi\left(\gamma + \delta g(y)\right)$$

on the support of Y and 0 elsewhere. This formula can be derived by the transformation technique discussed in Sect. 8.1. The random variable $X = \xi + \lambda Y$ is simply a linear transformation of Y, so that apart from location and scale, the shape of the pdf of X is the same as that of the pdf of Y. In other words, the shape of the pdf of X is essentially determined by the two parameters γ and δ and the specific function $g(\cdot)$.

6.2.1   Three Types of Distribution

Johnson [35] considered three different types of distribution of a random variable X, related to the three logarithmic functions

$$g(y) = \ln(y + \sqrt{1 + y^2}) ,$$

---

[1]That is, the set of values for which the pdf of Y is strictly positive.

$$g(y) = \ln(y/(1-y)), \quad 0 < y < 1 \,,$$

and

$$g(y) = \ln(y), \quad 0 < y \,.$$

The three types of distributions are called S_U (unbounded system), S_B (bounded system), and S_L (lognormal system). The corresponding pdfs of Y and X are given subsequently.

The S_U Distribution

Suppose that the variable $Y = (X - \xi)/\lambda$ has unbounded support and consider the function

$$g(Y) = \ln(Y + \sqrt{1 + Y^2}) = \sinh^{-1}(Y) \,.$$

Then, if $Z = \gamma + \delta g(Y)$ has standard normal distribution, the pdf of Y is

$$f(y) = \frac{\delta}{\sqrt{1 + y^2}} \, \phi\left(\gamma + \delta \sinh^{-1}(y)\right) \,,$$

and the pdf of $X = \xi + \lambda Y$ is given as

$$f(x) = \frac{\delta}{\sqrt{\lambda^2 + (x - \xi)^2}} \, \phi\left(\gamma + \delta \sinh^{-1}\left(\frac{x - \xi}{\lambda}\right)\right) \,.$$

The corresponding distribution is known as Johnson's S_U *distribution*. If X has S_U distribution, then

$$E(X) = \xi - \lambda\sqrt{e^{\delta^{-2}}} \sinh(\gamma/\delta)$$

and

$$\mathrm{Var}(X) = \frac{1}{2}\lambda^2(e^{\delta^{-2}} - 1)(e^{\delta^{-2}} \cosh(2\gamma/\delta) + 1) \,.$$

For $\gamma = 0$, the S_U distribution is symmetric about ξ. See Fig. 6.1 for the pdf of the S_U distribution for different choices of γ and δ.

The S_B Distribution

Suppose that the variable $Y = (X - \xi)/\lambda$ has the interval $(0, 1)$ as support and consider the function

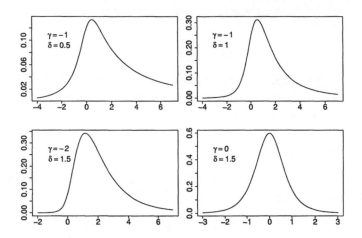

Fig. 6.1: Probability density function of the S_U distribution for $\xi = 0$, $\lambda = 1$ and different values of γ and δ

$$g(Y) = \ln(Y/(1 - Y)) = 2 \tanh^{-1}(2Y - 1) \ .$$

Then, if $Z = \gamma + \delta g(Y)$ has standard normal distribution, the pdf of Y is

$$f(y) = \frac{\delta}{y(1 - y)} \, \phi \left(\gamma + \delta \ln \left(\frac{y}{1 - y} \right) \right) I_{(0,1)}(y) \ ,$$

and the pdf of $X = \xi + \lambda Y$ is given as

$$f(x) = \frac{\delta \lambda}{(x - \xi)(\lambda - (x - \xi))} \, \phi \left(\gamma + \delta \ln \left(\frac{x - \xi}{\lambda - (x - \xi)} \right) \right) I_{(\xi, \xi + \lambda)}(x) \ .$$

The corresponding distribution is known as Johnson's S_B *distribution.*

The k-th non-central moment of Y can be expressed as

$$E(Y^k) = \frac{1}{\sqrt{2\pi}} \int_{-\infty}^{\infty} e^{-z^2/2} [1 + e^{-(z - \gamma)/\delta}]^{-k} \, dz \ ,$$

and must be evaluated numerically for given γ and δ. Johnson [35, p. 173 ff] also gives a rather complicated formula for $E(Y)$ not involving an integral and discusses the computation of the moments of Y. Having

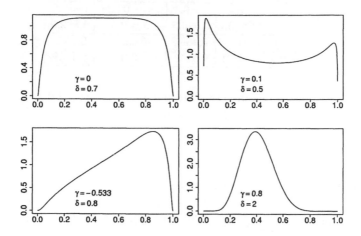

Fig. 6.2: Probability density function of the S_B distribution for $\xi = 0$, $\lambda = 1$ and different values of γ and δ

computed $E(Y)$ and $E(Y^2)$, it is easy to obtain expectation and variance of $X = \xi + \lambda Y$ for given ξ and λ.

As an example consider expectation and standard deviation of Y given in Table 6.1 for specific parameter configurations γ and δ, as also used in Fig. 6.2. For $\gamma = 0$ the S_B distribution is symmetric about 0.5 and thus has expectation 0.5.

The S_L Distribution

Suppose that the variable $Y = (X - \xi)/\lambda$ has the interval $(0, \infty)$ as support and consider the function

$$g(Y) = \ln(Y) .$$

Then, if the random variable $Z = \gamma + \delta g(Y)$ has standard normal distribution, the pdf of Y is

$$f(y) = \frac{\delta}{y} \, \phi \left(\gamma + \delta \ln(y) \right) I_{(0,\infty)}(y) ,$$

	$\gamma = 0,$ $\delta = 0.7$	$\gamma = 0.1,$ $\delta = 0.5$	$\gamma = -0.533,$ $\delta = 0.8$	$\gamma = 0.8,$ $\delta = 2.0$
$E(Y)$	0.5	0.4697	0.6251	0.4066
$\sqrt{\text{Var}(Y)}$	0.2631	0.3132	0.2312	0.1139

Table 6.1: Expectation and standard deviation of the S_B distribution for $\xi = 0$, $\lambda = 1$ and different values of γ and δ

and the pdf of $X = \xi + \lambda Y$ is given as

$$f(x) = \frac{\delta}{(x - \xi)} \, \phi \left((\gamma - \delta \ln(\lambda)) + \delta \ln(x - \xi) \right) I_{(\xi, \infty)}(x) \, .$$

Since for any set of parameter values $(\xi, \gamma, \delta, \lambda)$, it is possible to find a set of parameter values (ξ, γ, δ) such that the pdf

$$f(x) = \frac{\delta}{(x - \xi)} \, \phi \left(\gamma + \delta \ln(x - \xi) \right) I_{(\xi, \infty)}(x) \, .$$

is identical to the former one, this three-parameter pdf is usually studied. The corresponding distribution is called the S_L *distribution*. If X has S_L distribution, then

$$E(X) = \xi + e^{\delta^{-2} - 2\gamma\delta^{-1}} \quad \text{and} \quad \text{Var}(X) = e^{\delta^{-2} - 2\gamma\delta^{-1}} (e^{\delta^{-2}} - 1) \, .$$

The S_L distribution is skewed to the right, compare Fig. 6.3. In principle, it is also possible to apply a corresponding skewed to the left S_L distribution to given data, see Example 6.1, p. 155.

6.2.2   Estimation of Parameters

Slifker and Shapiro [77] develop a simple procedure to estimate the parameters of the Johnson distributions S_U, S_B and S_L. This procedure is based on four quantiles[2] obtained from the given observations $x_1, \ldots, x_n$. There are several slightly different methods known for the actual computation of

---

[2]An alternative estimation procedure based on quantiles is described by Wheeler [90].

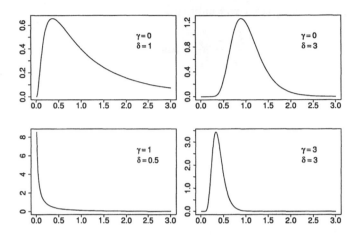

Fig. 6.3: Probability density function of the S_L distribution for $\xi = 0$ and different values of γ and δ

a p-quantile $\widehat{\xi}_p$ for $p \in (0, 1)$. Given the ordered values $x_{(1)}, \ldots, x_{(n)}$, one possible way is

$$\widehat{\xi}_p = (1 - \alpha)x_{(i)} + \alpha x_{(i+1)} ,$$

where i is the integer part of $(n - 1)p + 1$ and $\alpha = (n - 1)p + 1 - i$. This slightly differs from the proposed computation of quantiles given by Slifker and Shapiro.

Prerequisites

The estimation of the parameters requires the computation of four quantiles.

- As the first step, the user has to fix a positive number z. Slifker and Shapiro recommend the choice $z = 0.524$.
- As the second step, the probabilities $p_1 = \Phi(-3z)$, $p_2 = \Phi(-z)$, $p_3 = \Phi(z)$ and $p_4 = \Phi(3z)$ are computed.
- As the third step, the four quantiles $\widehat{\xi}_{p_1}$, $\widehat{\xi}_{p_2}$, $\widehat{\xi}_{p_3}$ and $\widehat{\xi}_{p_4}$ are computed.

- A the final step, the two quantities

$$\widetilde{x}_1 = \frac{\widehat{\xi}_{p4} - \widehat{\xi}_{p3}}{\widehat{\xi}_{p3} - \widehat{\xi}_{p2}} \quad \text{and} \quad \widetilde{x}_2 = \frac{\widehat{\xi}_{p2} - \widehat{\xi}_{p1}}{\widehat{\xi}_{p3} - \widehat{\xi}_{p2}}$$

are computed.

Discrimination

If one is not sure which of the three Johnson distributions S_U, S_B and S_L to fit to the given data, the above quantities may help to discriminate between them. For theoretical reasons, see [77], one may

$$\text{fit the} \left\{ \begin{array}{c} S_U \\ S_B \\ S_L \end{array} \right\} \text{distribution, if the product } \widetilde{x}_1 \widetilde{x}_2 \left\{ \begin{array}{cc} > & 1+\varepsilon \\ < & 1-\varepsilon \\ \in [1-\varepsilon, 1+\varepsilon] \end{array} \right\},$$

where ε is some small number, specified by the user, e.g. $\varepsilon = 0.1$.

The automatic discrimination, however, must be regarded with caution. It depends on the choice of z and may yield a decision in favor of the S_B (S_U) distribution, even if the true distribution of the sample is unbounded (bounded). See e.g. [14, 85] for further discussion.

The Estimates

Based on the value of $\widetilde{x}_1 \widetilde{x}_2$ being greater or smaller than 1, the given estimates by Slifker and Shapiro [77] either allow the application of the S_U or the S_B distribution. The estimates for the S_L distribution require $\widetilde{x}_1 > 1$, but see also Example 6.1 for the situation $\widetilde{x}_1 < 1$.

The S_U Distribution

It is assumed that $\widetilde{x}_1 \widetilde{x}_2 > 1$. Then the estimates for γ and δ are

$$\widehat{\gamma} = \widehat{\delta} \sinh^{-1} \left[\frac{\widetilde{x}_2 - \widetilde{x}_1}{2\sqrt{\widetilde{x}_1 \widetilde{x}_2 - 1}} \right] \quad \text{and} \quad \widehat{\delta} = \frac{2z}{\cosh^{-1}[(\widetilde{x}_1 + \widetilde{x}_2)/2]} .$$

The estimates for ξ and λ are

$$\widehat{\xi} = \frac{\widehat{\xi}_{p3} + \widehat{\xi}_{p2}}{2} + \frac{(\widehat{\xi}_{p3} - \widehat{\xi}_{p2})(\widetilde{x}_2 - \widetilde{x}_1)}{2(\widetilde{x}_1 + \widetilde{x}_2 - 2)}$$

and

$$\widehat{\lambda} = \frac{2(\widehat{\xi}_{p_3} - \widehat{\xi}_{p_2})\sqrt{\widetilde{x}_1\widetilde{x}_2 - 1}}{(\widetilde{x}_1 + \widetilde{x}_2 - 2)\sqrt{\widetilde{x}_1 + \widetilde{x}_2 + 2}} \, .$$

The S_B Distribution

It is assumed that $\widetilde{x}_1\widetilde{x}_2 < 1$. Let $\widetilde{y}_1 = 1/\widetilde{x}_1$ and $\widetilde{y}_2 = 1/\widetilde{x}_2$. Then the estimates for γ and δ are

$$\widehat{\gamma} = \widehat{\delta} \sinh^{-1}\left[\frac{(\widetilde{y}_2 - \widetilde{y}_1)\sqrt{(1 + \widetilde{y}_1)(1 + \widetilde{y}_2) - 4}}{2(\widetilde{y}_1\widetilde{y}_2 - 1)}\right]$$

and

$$\widehat{\delta} = \frac{z}{\cosh^{-1}\left[\sqrt{(1 + \widetilde{y}_1)(1 + \widetilde{y}_2)}/2\right]} \, .$$

The estimates for ξ and λ are

$$\widehat{\xi} = \frac{\widehat{\xi}_{p_3} + \widehat{\xi}_{p_2}}{2} - \frac{\widehat{\lambda}}{2} + \frac{(\widehat{\xi}_{p_3} - \widehat{\xi}_{p_2})(\widetilde{y}_2 - \widetilde{y}_1)}{2(\widetilde{y}_1\widetilde{y}_2 - 1)}$$

and

$$\widehat{\lambda} = \frac{(\widehat{\xi}_{p_3} - \widehat{\xi}_{p_2})\sqrt{((1 + \widetilde{y}_1)(1 + \widetilde{y}_2) - 2)^2 - 4}}{\widetilde{y}_1\widetilde{y}_2 - 1} \, .$$

The S_L Distribution

It is assumed that $\widetilde{x}_1 > 1$. Then the estimates for γ and δ are

$$\widehat{\gamma} = \widehat{\delta} \ln\left[\frac{\widetilde{x}_1 - 1}{(\widehat{\xi}_{p_3} - \widehat{\xi}_{p_2})\sqrt{\widetilde{x}_1}}\right] \quad \text{and} \quad \widehat{\delta} = \frac{2z}{\ln(\widetilde{x}_1)} \, .$$

The estimate for ξ is

$$\widehat{\xi} = \frac{\widehat{\xi}_{p_3} + \widehat{\xi}_{p_2}}{2} - \frac{(\widehat{\xi}_{p_3} - \widehat{\xi}_{p_2})}{2}\frac{\widetilde{x}_1 + 1}{\widetilde{x}_1 - 1} \, .$$

Example 6.1 Consider the female athletes weights data from Table 1.1. Suppose that we wish to fit a Johnson distribution. We start the estimation procedure by fixing the value $z = 0.524$, giving the four p_i-quantiles in Table 6.2.

i	1	2	3	4
p_i	0.0580	0.3001	0.6999	0.9420
$\widehat{\xi}_{p_i}$	48.84	62.73	74.10	82.83

Table 6.2: Four p_i-quantiles for the data from Table 1.1

Then $\widetilde{x}_1 = 0.7673$ and $\widetilde{x}_2 = 1.2217$, so that the product $\widetilde{x}_1\widetilde{x}_2 = 0.9375$ is smaller than 1, indicating that the S_B distribution should be appropriate. On the other hand, the value is not far away from 1, implying that also the S_L distribution may be fitted. In the following we fit both distributions, starting with S_B. It is straightforward to compute the parameter estimates for the S_B distribution; the values are given in Table 6.3. As it turns out, the S_B fit has support $(\widehat{\xi}, \widehat{\xi} + \widehat{\lambda}) = (-48.6293, 102.8273)$.

$\widehat{\gamma}$	$\widehat{\delta}$	$\widehat{\xi}$	$\widehat{\lambda}$
-3.0091	2.4330	-48.6293	151.4566

Table 6.3: Parameter estimates of the S_B distribution for the data from Table 1.1

The fit of S_L distribution is a little involved, since the S_L distribution is skewed to the right, while the data is skewed to the left. Not surprisingly, the above estimation procedure fails, since in view of $\widetilde{x}_1 < 1$ the estimate of γ does not exist. Using a little trick, however, makes it possible to apply the above theory for obtaining a skewed to the left S_L fit. Simply consider the original data points multiplied by -1, the distribution of which being skewed to the right, estimate the density of the S_L distribution as $f(x; \widehat{\gamma}, \widehat{\xi}, \widehat{\lambda})$ from the new data, and let $f(-x; \widehat{\gamma}, \widehat{\xi}, \widehat{\lambda})$ be the pdf fitting the original data, having support $(-\infty, -\widehat{\xi})$. For the female athletes weights data, this yields a fit by the pdf

$$f(x) = \frac{\widehat{\delta}}{(-x - \widehat{\xi})} \, \phi\left(\widehat{\gamma} + \widehat{\delta}\ln(-x - \widehat{\xi})\right) I_{(-\infty, -\widehat{\xi})}(x) ,$$

where $\widehat{\delta} = 5.2335$, $\widehat{\gamma} = -21.1312$, and $\widehat{\xi} = -125.3921$.

The S_U distribution cannot be fitted to the data , since $\widetilde{x}_1 \widetilde{x}_2 < 1$. But $\widetilde{x}_1$ and $\widetilde{x}_2$ depend on the choice of z, and it may be the case that a different choice of z will give $\widetilde{x}_1 \widetilde{x}_2 > 1$. As a matter of fact, $\widetilde{x}_1 \widetilde{x}_2 = 1.0580$ for the choice $z = 0.7$.

$\widehat{\gamma}$	$\widehat{\delta}$	$\widehat{\xi}$	$\widehat{\lambda}$
0.1038	5.8498	71.8613	60.3466

Table 6.4: Parameter estimates of the S_U distribution for the data from Table 1.1

Then, the estimation of the S_U distribution parameters yields the values given in Table 6.4. Note that the estimates are *not* based on the four quantiles from Table 6.2, but on four quantiles obtained from the choice $z = 0.7$. Fig. 6.4 shows the fitted S_B, S_L and S_U distributions. □

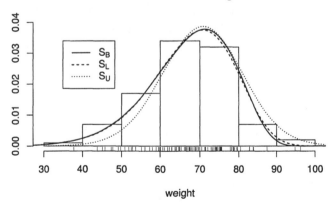

Fig. 6.4: Fit of Johnson distributions to the data from Table 1.1

6.2.3   Choice of Fit

In Example 6.1 we have shown how to fit each of the Johnson distributions S_B, S_L and S_U to a given data set. In order to decide which of the fits to use, one may apply the 'starship' procedure by Owen [59], see also [71, Sect. 5]. Since the random variable $Z = \widehat{\gamma} + \widehat{\delta} g\left(\frac{X-\widehat{\xi}}{\widehat{\lambda}}\right)$ should follow a normal distribution, simply choose the set of estimates which admits the highest indication of normality. In order to check this, a normality test may be applied to the observations $z_1, \ldots, z_n$ of Z, obtained from the original data points $x_1, \ldots, x_n$ by the above transformation.

This procedure can also be used in general to find the value of z (i.e. the number to be fixed by the user in the Slifker and Shapiro estimation procedure), being the value which yields the highest indication of normality. See [60] for an implementation of the starship procedure. When this is applied, the value of z must no longer be fixed by the user, but is *found by the data*.

When we consider the female athletes weights data, and the parameter estimates given in Example 6.1 for the S_B, S_L and S_U distributions, then normal quantile-quantile plots[3] of the respective transformed samples $z_1, \ldots, z_n$ show that the S_U distribution fit gives the best approximation to the normal distribution, see Fig. 6.5. Different normality tests as described in Sect. 4 confirm this conclusion. Hence, from this point of view, the S_U distribution fit obtained from the choice $z = 0.7$ is preferable to the S_B and S_L fits. As noted in the previous paragraph, one may of course check further choices of z, possibly admitting an even higher indication of normality.

6.3   The Lognormal Distribution

Often, a real-world random variable X can only take nonnegative values, and in addition the distribution of these values is seen to be skewed to the right. In such an instance, it is widespread practice to consider the logarithm of the observations as new sample values and then fit the normal distribution to this new sample. This can be justified when a normal

---

[3]For the S_L distribution the negative female athletes weights have been used for obtaining the transformed sample with $\widehat{\lambda} \equiv 1$.

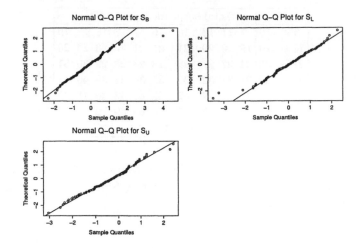

Fig. 6.5: Normal quantile-quantile plots of the data points obtained from Johnson's S_B, S_L and S_U transformations on the female athletes weights data points given in Table 1.1

quantile-quantile plot of the logarithms reveals a good indication of normality[4].

As an example, consider the data set given in Table 6.5. A histogram of the data shows that the distribution of the nonnegative values is skewed to the right, see Fig 6.6 (a). The normal quantile-quantile plot of the logarithms reveals a good indication of normality, see Fig. 6.6 (b). In addition, different normality tests do not reject the hypothesis of normality. For example, the value of the Shapiro-Francia test statistic comes to $W' = 0.9888$, yielding a p-value of 0.7359.

---

[4]On the other hand, there may also exist distributions (e.g. the gamma distribution) which are even more appropriate.

	Total suspended particulates									
1 – 10	7	11	13	16	16	17	17	18	18	19
11 – 20	19	20	20	21	21	21	22	23	24	24
21 – 30	24	27	27	28	28	28	28	29	30	31
31 – 40	31	31	32	33	35	36	37	39	40	41
41 – 50	41	41	42	42	42	42	42	46	46	47
51 – 60	47	51	55	64	64	65	69	72	74	74
61 – 65	81	88	91	101	124					

Table 6.5: Total suspended particulate measurements ($\mu g/m^3$) at an air sampling station near the Navajo Generating Station, 1974, as given in [86, p. 340]

6.3.1 Two- and Three-Parameter Lognormal Distribution

If X is a random variable having the interval $(0, \infty)$ as support, then X is said to have a *lognormal distribution* with parameters μ and σ^2 if $\ln(X)$ has a normal distribution with parameters μ and σ^2. The pdf of X is then given as

$$f(x) = \frac{1}{\sqrt{2\pi}\sigma x} e^{-\frac{1}{2}(\ln(x)-\mu)^2/\sigma^2} I_{(0,\infty)}(x) .$$

This definition may easily be extended to a random variable X having the interval (ξ, ∞) as support, yielding the pdf

$$f(x) = \frac{1}{\sqrt{2\pi}\sigma(x-\xi)} e^{-\frac{1}{2}(\ln(x-\xi)-\mu)^2/\sigma^2} I_{(\xi,\infty)}(x) .$$

As can be seen, this is the pdf of the S_L distribution with

$$\gamma = -\frac{\mu}{\sigma} \quad \text{and} \quad \delta = \frac{1}{\sigma} .$$

Often, it is assumed that the value of ξ is known, in which case the distribution only depends on two parameters instead of three.

Definition 6.1 *Let X be a random variable having the interval (ξ, ∞) as support. When $\ln(X-\xi)$ has normal distribution with parameters μ and σ^2,*

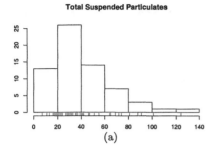
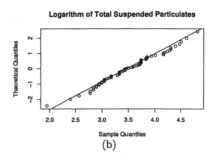

Fig. 6.6: Histogram of data from Table 6.5 (a) and normal quantile-quantile plot of corresponding logarithmic values (b)

X is said to have a two-parameter lognormal distribution *with parameters* μ *and* σ^2 *if* ξ *is known, and a* three-parameter lognormal distribution *with parameters* ξ, μ *and* σ^2, *if* ξ *is unknown.*

If X has a (two- or three-) parameter lognormal distribution, then

$$\mathrm{E}(X) = \xi + e^{\sigma^2/2+\mu} \quad \text{and} \quad \mathrm{Var}(X) = e^{\sigma^2+2\mu}(e^{\sigma^2} - 1) .$$

See also [2, 17, 36] for further reading on the lognormal distribution. Aitchison and Brown [2, Chapter 10] give examples for real-life variables whose observed distribution has been fitted by a lognormal distribution.

6.3.2 Estimation of Parameters

When ξ is known, then instead of the original sample $X_1, \ldots, X_n$, one may consider the transformed sample $Z_1, \ldots, Z_n$, $Z_i = \ln(X_i - \xi)$, which can be assumed to com from a normal distribution with parameters μ and σ^2. These parameters may then be estimated by

$$\widehat{\mu} = \overline{Z} \quad \text{and} \quad \widehat{\sigma}^2 = \frac{1}{n-1} \sum_{i=1}^{n} (Z_i - \overline{Z})^2 .$$

If ξ is not known, then one possibility is to apply the method of Slifker and Shapiro to the parameters of the S_L distribution.

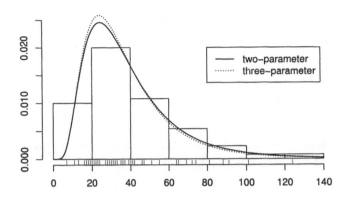

Fig. 6.7: Two- and three-parameter lognormal fit to the data from Table 6.5

Example 6.2 Consider the data from Table 6.5 for which we want to fit a two- and a three-parameter lognormal distribution. Let us first consider the two-parameter case for $\xi = 0$. Then the estimates obtained from the logarithmic values are $\hat{\mu} = 3.5121$ and $\hat{\sigma}^2 = 0.3236$. For the three-parameter case we apply the method by Slifker and Shapiro. For the choice of z, a variant of the starship procedure is employed, choosing that value of z between 0.4 and 0.9 (with a precision of three digits after the dot) which yields the greatest value of the Shapiro-Francia test statistic[5]. As it turns out, there are choices for z which yield an estimate $\hat{\xi}$ being greater than 7 or smaller than 0. However, since the smallest data point is 7 and values smaller than zero cannot occur, it is reasonable to confine only to those values of z yielding an estimate $0 \leq \hat{\xi} < 7$. Among those, the choice $z = 0.625$ yields the greatest value of the Shapiro-Francia statistic, being 0.9888. (Recall from above that this is also the value of the Shapiro-Francia statistic when simply the logarithmic values are used.) Then the

---

[5]Recall from Sect. 4.4.1 that the Shapiro-Francia test statistic W' takes values between 0 and 1, greater values indicating a higher degree of normality.

estimates of the S_L distribution parameters are $\widehat{\xi} = 0.0071$, $\widehat{\gamma} = -6.3041$, and $\widehat{\delta} = 1.8108$. From these, $\widehat{\mu} = -\widehat{\gamma}/\widehat{\delta} = 3.4813$ and $\widehat{\sigma}^2 = 0.3050$, showing that the so obtained three-parameter fit is close to the two-parameter fit, see Fig. 6.7. □

6.4 Power Transformations

The concept of power transformations of a given set of observations has mainly been investigated in connection with linear models, see e.g. [5]. Such transformations may also prove helpful when it is desired to transform a single random variable to normality. For this, it is assumed that X can only take nonnegative values, or that there is some user specified ξ such that $X + \xi$ can only take nonnegative values and the corresponding sample $x_1 + \xi, \ldots, x_n + \xi$ is regarded.

6.4.1 Box-Cox Transformation

A generalization of the logarithmic transformation $\ln(X)$ of a random variable X is the *Box-Cox power transformation* [10], being

$$Z(\lambda) = \begin{cases} \dfrac{X^\lambda - 1}{\lambda} & \text{if } \lambda \neq 0 \\ \ln(X) & \text{if } \lambda = 0 \end{cases}.$$

The case $\lambda = 0$ is obtained from the case $\lambda \neq 0$ with $\lambda \to 0$. As a matter of fact, since $\mathrm{d}(X^\lambda - 1)/\mathrm{d}\lambda = X^\lambda \ln(X)$, by the rule of l'Hopital

$$\lim_{\lambda \to 0} \frac{X^\lambda - 1}{\lambda} = \lim_{\lambda \to 0} X^\lambda \ln(X) = \ln(X).$$

Gaudard and Karson [26] compare methods for obtaining a data based choice of λ. A quite natural and easy to apply method is simply to choose that value of λ for which the transformed data admits the highest indication of normality. This may be measured by a normality test statistic, e.g. the Shapiro-Francia statistic.

Example 6.3 Consider $n = 100$ data points $x_1, \ldots, x_n$ as shown in Fig. 6.8 (a). Clearly, the distribution of the data is not normal, being also reflected by corresponding normal quantile-quantile plot in Fig.

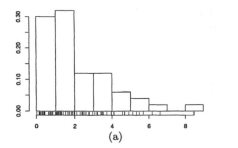 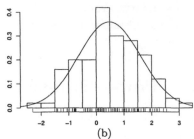

(a) (b)

Fig. 6.8: Histogram of (a) $n = 100$ data points, and (b) of the Box-Cox transformed values with $\lambda = 0.3$, superimposed by estimated normal pdf

6.9 (a). For each $\lambda = -2.00, -1.95, \ldots, 1.95, 2.00$ a transformed sample $z_1(\lambda), \ldots, z_n(\lambda)$ with

$$z_i(\lambda) = \begin{cases} \dfrac{x_i^\lambda - 1}{\lambda} & \text{if } \lambda \neq 0 \\ \ln(x_i) & \text{if } \lambda = 0 \end{cases},$$

is generated, from which the corresponding Shapiro-Francia test statistic W' is computed. Among these, the choice $\lambda = 0.3$ yields the maximal value $W' = 0.9931$ with a corresponding p-value of 0.824. Fig. 6.8 (b) shows the Box-Cox transformed data points, superimposed by the estimated normal pdf. Although the choice $\lambda = 0.3$ is optimal within the considered set, the transformation to normality need not necessarily be perfect, as can also be seen from Fig. 6.9 (b). □

Note that the considered range of λ values from -2 to 2 in the above example is arbitrary, but smaller or greater values are not expected to be of relevance for the given data. This might of course be different for other samples, for which the considered range can be chosen differently by the investigator.

When the same procedure as in Example 6.3 is carried out for the data from Table 6.5, it turns out that in fact the choice $\lambda = 0$ is optimal, thus additionally justifying the fit of the lognormal distribution to the data.

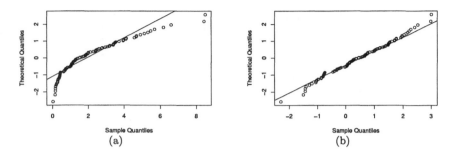

Fig. 6.9: Normal quantile-quantile plot (a) of the $n = 100$ data points from Fig. 6.8, and (b) of the Box-Cox transformed values with $\lambda = 0.3$

In general, if a value λ near zero is obtained, one should check whether putting λ equal to zero will give almost the same indication of normality. If so, additional advantage can be drawn from the fit of a known distribution, namely the lognormal distribution.

6.4.2 Transformation for Proportions

Quite often, a given data set consists of percentages or proportions, where percentages may also be handled as proportions when divided by 100. For such data, it is seen that the Box-Cox transformation often fails to admit a satisfactory indication of normality.

Folded Power Transformation

As an alternative transformation of a random variable X with support $[0, 1]$, one may consider the so-called *folded power transformation* [57]

$$
Z(\lambda) = \begin{cases} \dfrac{X^\lambda - (1 - X)^\lambda}{\lambda} & \text{if } \lambda \neq 0 \\ \ln(X/(1 - X)) & \text{if } \lambda = 0 \end{cases} .
$$

Similarly to the Box-Cox transformation, the case $\lambda = 0$ is obtained from the case $\lambda \neq 0$ but $\lambda \to 0$.

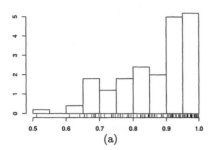 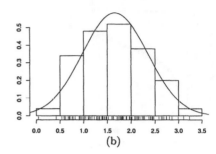

(a)　　　　　　　　　　　　　　　　(b)

Fig. 6.10: Histogram of $n = 100$ proportion data points (a), and histogram of Guerrero and Johnson transformed values with $\lambda = -0.25$, superimposed by the estimated normal pdf (b)

Guerrero and Johnson Transformation

As an alternative to the folded power transformation applied to proportions, one may also consider the proposal by Guerrero and Johnson [29] to use the Box-Cox transformation not on X, but on $X/(1 - X)$, yielding the transformation

$$Z(\lambda) = \begin{cases} \dfrac{(X/(1-X))^\lambda - 1}{\lambda} & \text{if } \lambda \neq 0 \\ \ln(X/(1-X)) & \text{if } \lambda = 0 \end{cases}.$$

As opposed to the folded power transformation, this transformation has the advantage that for known λ the original value can immediately be computed from the transformed value. As a matter of fact, if x is an observation transformed to z by

$$z = \frac{(x/(1-x))^\lambda - 1}{\lambda},$$

then

$$x = \frac{(\lambda z + 1)^{1/\lambda}}{(\lambda z + 1)^{1/\lambda} + 1}.$$

Moreover, when λ is chosen in the same way as described before, the Guerrero and Johnson transformation may even admit a higher indication of normality than the folded power transformation. Nonetheless, it is not guaranteed that any of these transformations will admit a satisfyingly indication of normality.

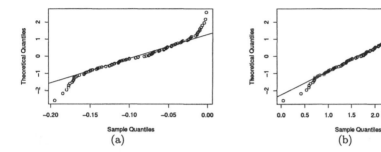

Fig. 6.11: Normal quantile-quantile plot of the Box-Cox transformed data from Example 6.4 with $\lambda = 4.95$ (a), and normal quantile-quantile plot of the Guerrero and Johnson transformed data with $\lambda = -0.25$ (b)

Example 6.4 Consider $n = 100$ data points $x_1, \ldots, x_n$ as shown in Fig. 6.10 (a). The data points[6] are known to lie between 0 and 1. The Box-Cox parameter $\lambda = 4.95$ admits the best indication of normality with respect to the Shapiro-Francia test statistic, being $W' = 0.9546$ with corresponding p-value of 0.0025 for $\lambda = 4.95$. As can also be seen from the normal quantile-quantile plot of the transformed data in Fig. 6.11 (a), the Box-Cox transformation to normality is not successful. When instead we consider the folded power transformation, then the choice $\lambda = 0.25$ yields a value $W' = 0.9905$ with corresponding p-value 0.6136. For the choice $\lambda = -0.25$, the Guerrero and Johnson transformation yields $W' = 0.9922$ with corresponding p-value 0.7476. In addition, the normal quantile-quantile plot of the Guerrero and Johnson transformed data in Fig. 6.11 (b) gives a good but not perfect indication of normality. Fig. 6.10 (b) shows a histogram of the Guerrero and Johnson transformed data with superimposed estimated normal pdf. □

---

[6]Actually, the data has been generated from a beta distribution with parameters $p = 6$ and $q = 1$.

Chapter 7

Two Normal Variables

Often, given data obtained from n units of measurement does not only consist of observations from one but several random variables at the same time. In this chapter we confine to two normal variables X and Y admitting pairs of observations $(x_1, y_1), \ldots, (x_n, y_n)$, each pair being the observation from an individual unit of measurement. Being interested in the distribution of X and Y, we distinguish between two cases:

- X and Y having a joint normal distribution;
- X and Y both having a normal distribution.

On first sight, there seems to be no difference, but as a matter of fact there is. In the first case we are concerned with the distribution of the two-dimensional random variable (X, Y), while in the second case we are concerned with the individual distributions of X and Y. Although there are relationships between both cases, they are not the same.

7.1 The Bivariate Normal Distribution

When the joint distribution of k random variables is of interest, the k-*variate normal distribution* is a natural generalization of the normal distribution and is of similar importance. In this text we confine to the case $k = 2$, but see e.g. [88] for a treatment of the k-variate normal distribution.

7.1.1   Probability Density Function

Usually the bivariate normal distribution is defined via its probability density function.

Definition 7.1 *The two-dimensional random variable* (X, Y) *is said to have a* bivariate normal distribution *with parameters* μ_1, μ_2, $\sigma_1 > 0$, $\sigma_2 > 0$ *and* $-1 < \varrho < 1$, *if the joint pdf of* X *and* Y *is given as*

$$f(x, y) = \frac{1}{2\pi\sigma_1\sigma_2\sqrt{1 - \varrho^2}}\, \mathrm{e}^{-\frac{1}{2}Q(x,y;\mu_1,\mu_2,\sigma_1,\sigma_2,\varrho)} ,$$

where

$$\begin{aligned}
&Q(x, y; \mu_1, \mu_2, \sigma_1, \sigma_2, \varrho) \\
&= \frac{1}{(1 - \varrho^2)}\left[\frac{(x - \mu_1)^2}{\sigma_1^2} - 2\varrho\frac{(x - \mu_1)(y - \mu_2)}{\sigma_1\sigma_2} + \frac{(y - \mu_2)^2}{\sigma_2^2}\right] .
\end{aligned}$$

See Fig. 7.1 for a representation of the bivariate normal pdf. When (X, Y) has a bivariate normal distribution, then X and Y are also said to have a *joint normal distribution*.

Given parameters $\mu_1, \mu_2, \sigma_1, \sigma_2$, ϱ, and a fixed c, the bivariate normal pdf $f(x, y)$ is the same for any (x, y) from the set of points

$$\{(x, y):\ Q(x, y; \cdots) = c^2\} .$$

These sets (for different values c^2) are also called the *contours* of the bivariate normal distribution. They describe ellipses in the xy-plane with its center in (μ_1, μ_2). If $\varrho = 0$, the axes of the ellipses are parallel to x- and y-axis, respectively. If in addition $\sigma_1 = \sigma_2$, both axes have equal length, so that the ellipses become circles, see Fig. 7.2.

7.1.2   Some Properties

The parameters of the bivariate normal distribution have an interpretation with respect to the distribution of individual variables X and Y, μ_1 and μ_2 being the expectations of X and Y, σ_1^2 and σ_2^2 being the variances of X and Y, and ϱ being the correlation between X and Y.

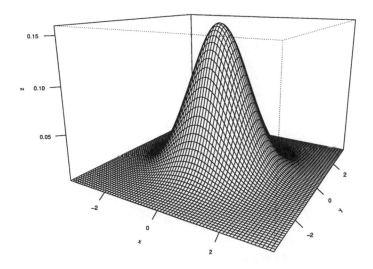

Fig. 7.1: Probability density function of the bivariate normal distribution with $\mu_1 = 0$, $\mu_2 = 0$, $\sigma_1 = 1$, $\sigma_2 = 1$ and $\varrho = 0$

Theorem 7.1 *If the two-dimensional random variable (X, Y) has bivariate normal distribution with parameters μ_1, μ_2, $\sigma_1 > 0$, $\sigma_2 > 0$ and $-1 < \varrho < 1$, then*

(a) *X is normally distributed with expectation μ_1 and variance σ_1^2,*
(b) *Y is normally distributed with expectation μ_2 and variance σ_2^2,*
(c) $\varrho = \mathrm{Cov}(X, Y)/\sqrt{\mathrm{Var}(X)\mathrm{Var}(Y)}$.

The converse of statements (a) and (b) is not true in general. When X and Y have individual normal distributions, then (X, Y) does not necessarily have a bivariate normal distribution, compare Sect. 7.2.1 for more on this. Since we have $|\varrho| < 1$, statement (c) also implies that perfect correlation between X and Y is excluded.

When (X, Y) has a bivariate normal distribution, then the individual random variables X and Y are specific linear combinations of X and Y,

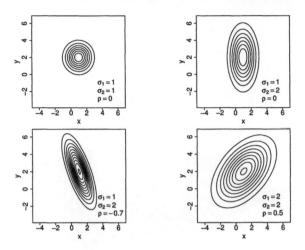

Fig. 7.2: Contour plots of the bivariate normal pdf with $\mu_1 = 1$, $\mu_2 = 2$ and different values of σ_1, σ_2 and ϱ

respectively, being normally distributed. The following theorem states that *every* linear combination of X and Y has again a normal distribution.

Theorem 7.2 *If the two-dimensional random variable (X, Y) has* bivariate normal distribution *with parameters μ_1, μ_2, $\sigma_1 > 0$, $\sigma_2 > 0$ and $-1 < \varrho < 1$, then the linear combination*

$$aX + bY$$

with a and b not both being zero, is normally distributed with expectation $a\mu_1 + b\mu_2$ and variance $a^2\sigma_1^2 + 2\varrho ab\sigma_1\sigma_2 + b^2\sigma_2^2$.

The converse of the above theorem is also true provided that X and Y are not perfectly correlated, i.e. $|\varrho_{X,Y}| < 1$. Under this assumption, if $aX + bY$ has a normal distribution for *every* a and b not both equal to zero, then (X, Y) has a bivariate normal distribution. This property may also be used to characterize the bivariate normal distribution.

Theorem 7.3 *A two-dimensional random variable (X, Y) with $|\varrho_{X,Y}| < 1$ has a bivariate normal distribution if and only if every non-trivial linear combination of X and Y has a univariate normal distribution.*

The above characterization may even be generalized to an arbitrary two-dimensional random variable (X, Y) with $|\varrho_{X,Y}|$ being possibly equal to 1, in which case, however, the joint pdf $f(x, y)$ cannot be given in the closed form as above. This is also called a *singular* bivariate normal distribution.

7.1.3   Uncorrelatedness and Independence

The pdf of the bivariate normal distribution may also be written in the form

$$f(x, y) = \frac{1}{\sigma_1 \sigma_2} \phi \left(\frac{x - \mu_1}{\sigma_1} \right) \frac{1}{\sqrt{1 - \varrho^2}} \phi \left(\frac{y - \mu_2}{\sigma_2 \sqrt{1 - \varrho^2}} - \varrho \frac{(x - \mu_1)}{\sigma_1 \sqrt{1 - \varrho^2}} \right) .$$

For the special case $\varrho = 0$, the joint pdf reduces to

$$f(x, y) = \frac{1}{\sigma_1 \sigma_2} \phi \left(\frac{x - \mu_1}{\sigma_1} \right) \phi \left(\frac{y - \mu_2}{\sigma_2} \right) ,$$

being simply the product of individual normal pdfs.

Theorem 7.4 *If the two-dimensional random variable (X, Y) has bivariate normal distribution with parameters μ_1, μ_2, $\sigma_1 > 0$, $\sigma_2 > 0$ and $-1 < \varrho < 1$, then the individual variables X and Y are independent if and only if they are uncorrelated.*

The 'only if' part of the above theorem is always true for two random variables X and Y, even if they do not have normal distributions. The 'if' part is true under the stated assumption of bivariate normality, but also other assumptions are possible, see Sect. 7.2.3. However, the often heard statement 'two normally distributed random variables are independent if and only if they are uncorrelated' is incorrect, see Sect. 7.2.2.

As noted above, the product of two individual normal pdfs is a special case of the bivariate pdf, implying that independent normal random variables also have a joint normal distribution.

Theorem 7.5 *If X and Y are two independent variables having normal distributions, then (X, Y) has a bivariate normal distribution.*

7.1.4   Estimation

When we have pairs of observations $(x_1, y_1), \ldots, (x_n, y_n)$ of a two-dimensional random variable (X, Y) being bivariate normally distributed, then obvious estimates of the parameters are

$$\widehat{\mu}_1 = \overline{x}, \quad \widehat{\mu}_2 = \overline{y},$$

$$\widehat{\sigma}_1^2 = \frac{1}{n-1} \sum_{i=1}^{n} (x_i - \overline{x})^2, \quad \widehat{\sigma}_2^2 = \frac{1}{n-1} \sum_{i=1}^{n} (y_i - \overline{y})^2,$$

and

$$\widehat{\varrho} = \frac{\displaystyle\sum_{i=1}^{n}(x_i - \overline{x})(y_i - \overline{y})}{\sqrt{\displaystyle\sum_{i=1}^{n}(x_i - \overline{x})^2 \sum_{i=1}^{n}(y_i - \overline{y})^2}}.$$

7.1.5   Assessing Bivariate Normality

It is not easy to asses bivariate normality of given data, since it is not enough to demonstrate that X and Y can be assumed to follow a normal distribution. Nonetheless normal quantile-quantile plots of the respective X and Y observations can be helpful for either dismissing bivariate normality or provoking further analysis. In addition, peculiarities in the data (like outliers) may also be found by a scatter plot of the (x_i, y_i) pairs. A further graphical tool can be a *beta quantile-quantile plot* of the *modified estimated radii*, where the estimated radii are given by

$$r_i^2 = Q(x_i, y_i; \widehat{\mu}_1, \widehat{\mu}_2, \widehat{\sigma}_1, \widehat{\sigma}_2, \widehat{\varrho})$$

for $i = 1, \ldots, n$. Then for theoretical reasons, see also [86, Sect. 9.2.2], the ordered values $nr_i^2/(n-1)^2$ can be plotted against the

$$F^{-1}\left(\frac{i}{n + (n-3)^{-1} + 0.5}\right)$$

values, where $F^{-1}(\cdot)$ is the quantile function of the beta distribution with parameters $p = 1$ and $q = (n - 3)/2$. See p. 204 for the pdf of the beta distribution.

No.	sepal length	sepal width	No.	sepal length	sepal width
1	5.1	3.5	26	5.0	3.0
2	4.9	3.0	27	5.0	3.4
3	4.7	3.2	28	5.2	3.5
4	4.6	3.1	29	5.2	3.4
5	5.0	3.6	30	4.7	3.2
6	5.4	3.9	31	4.8	3.1
7	4.6	3.4	32	5.4	3.4
8	5.0	3.4	33	5.2	4.1
9	4.4	2.9	34	5.5	4.2
10	4.9	3.1	35	4.9	3.1
11	5.4	3.7	36	5.0	3.2
12	4.8	3.4	37	5.5	3.5
13	4.8	3.0	38	4.9	3.6
14	4.3	3.0	39	4.4	3.0
15	5.8	4.0	40	5.1	3.4
16	5.7	4.4	41	5.0	3.5
17	5.4	3.9	42	4.5	2.3
18	5.1	3.5	43	4.4	3.2
19	5.7	3.8	44	5.0	3.5
20	5.1	3.8	45	5.1	3.8
21	5.4	3.4	46	4.8	3.0
22	5.1	3.7	47	5.1	3.8
23	4.6	3.6	48	4.6	3.2
24	5.1	3.3	49	5.3	3.7
25	4.8	3.4	50	5.0	3.3

Table 7.1: Sepal length and sepal width of 50 iris setosa flowers in centimeters

Example 7.1 Consider the observations of sepal length and sepal width in centimeters on $n = 50$ flowers from the iris setosa species as given in Table 7.1. This is a part of the famous Fisher's iris data set. Fig. 7.3 (a) shows a scatter plot of the observations with three of them marked by their number in the data set. Fig. 7.3 gives the corresponding beta quantile-quantile plot. As it seems, the observation with number 42 looks even more outstanding in the beta quantile-quantile plot than in the scatter plot. Such a behaviour might be worth taking into account when interpreting a beta quantile-quantile plot of modified estimated radii.   □

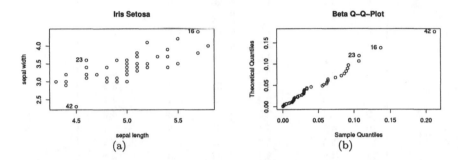

Fig. 7.3: Scatter plot of data pairs (a) and beta quantile-quantile plot of modified ordered radii (b) for the data from Table 7.1

See also [86, Sect. 9] for a review of more multivariate plots and tests possibly revealing departures from multivariate normality. Marden [46] gives a further method for bivariate quantile-quantile plots.

7.1.6   Random Sample Generation

To obtain a pseudo-random sample from the bivariate normal distribution, let z_1 and z_2 be two independent observations from the standard normal distribution. Then the pair (x, y) with

$$x = \sigma_1 z_1 + \mu_1$$

and

$$y = \sigma_2 \varrho z_1 + \sigma_2 \sqrt{1 - \varrho^2} z_2 + \mu_2$$

can be seen as an observation from the bivariate normal distribution with parameters μ_1, μ_2, σ_1, σ_2 and ϱ. Compare also the last paragraph in Sect. 7.3. Repeating independently n times the generation of pairs (z_1, z_2), and computing x and y as above, yields the desired sample, see Fig. 7.4 for an illustration.

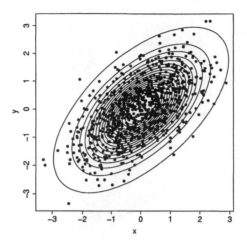

Fig. 7.4: 1000 simulated observations from the bivariate normal distribution with parameters $\mu_1 = \mu_2 = 0$, $\sigma_1 = \sigma_2 = 1$, and $\varrho = 0.6$

7.2   Two Normal Variables

When we consider two normally distributed random variables X and Y (not necessarily independent of each other), then the two-dimensional random variable (X, Y) does not necessarily have a bivariate normal distribution, as discussed in the following Sect. 7.2.1. Ignoring the difference between individual and joint normality may be seen as one of the main sources for the fallacies described subsequently to which we refer in form of questions often erroneously answered by 'yes'.

7.2.1   Are Two Normals Jointly Normal?

The answer to the question is *no, not in general*. It is possible to find examples of random variables X and Y being normally distributed, but not having the bivariate normal distribution as their joint distribution. One possible reason for the common fallacy that 'individual normality implies joint normality' might be fact that this statement is true in case that X and Y are independent.

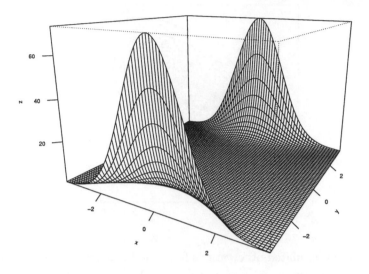

Fig. 7.5: Joint pdf of X and Y in Example 7.2

Example 7.2 (See also [69, Sect. 2.10].) Let the joint pdf of X and Y be given as

$$f(x,y) = \begin{cases} \frac{1}{2\pi}\,\mathrm{e}^{-\frac{1}{2}(x^2-y^2)} + \frac{1}{4\pi\mathrm{e}}x^3y^3 & \text{if } -1 \le x \le 1 \text{ and } -1 \le y \le 1 \\ \frac{1}{2\pi}\,\mathrm{e}^{-\frac{1}{2}(x^2-y^2)} & \text{otherwise} \end{cases},$$

see Fig. 7.5 for an illustration. Then the pdf of X is

$$f(x) = \int_{-\infty}^{\infty} f(x,y)\,\mathrm{d}y$$
$$= \int_{-\infty}^{\infty} \frac{1}{2\pi}\,\mathrm{e}^{-\frac{1}{2}(x-y)^2}\,\mathrm{d}y + \underbrace{\int_{-1}^{1} \frac{1}{4\pi\mathrm{e}}x^3y^3\,\mathrm{d}y}_{=0} = \frac{1}{\sqrt{2\pi}}\,\mathrm{e}^{-x^2/2},$$

and thus the pdf of the standard normal distribution. Similarly Y has standard normal distribution, but the joint pdf $f(x,y)$ is not the bivariate normal one. □

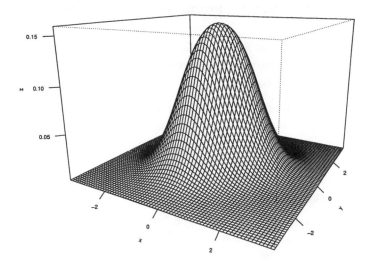

Fig. 7.6: Joint pdf of X and Y in Example 7.3

Example 7.3 (See also [63, 9.1.2].) Let

$$f(x,y) = \phi(x)\phi(y)\left[1 + \alpha(1 - 2\Phi(x))(1 - 2\Phi(y))\right]$$

for some α satisfying $|\alpha| < 1$. See Fig. 7.6 for an illustration with $\alpha = 0.9$. If $\alpha \neq 0$, then $f(x,y)$ is not the bivariate normal pdf, but the pdfs of X and Y are standard normal. $\square$

Example 7.4 (See also [12, Exercise 4.47].) Let X and Y be independent standard normal variables, and let Z be

$$Z = \begin{cases} X & \text{if } XY > 0 \\ -X & \text{otherwise} \end{cases}.$$

Then Z has standard normal distribution, but the joint pdf of the two normals Z and Y is not bivariate normal. (Z and Y always have the same sign.) $\square$

7.2.2   Are Two Uncorrelated Normals Independent?

The answer to the question is *no, not in general*. Although it is an often heard fallacy that 'two normals are uncorrelated if and only if they are independent', this is not correct.

It is a fact that independence of two random variables (not necessarily having normal distributions) always implies their uncorrelatedness, but the converse implication is not necessarily true. Numerous examples for two uncorrelated but dependent random variables can be found in the literature. However, examples for uncorrelated but dependent normals are quite rare, although existing.

The source of the fallacy might be the fact that it is often stated that 'two variables having joint normal distribution are uncorrelated if and only if they are independent', which is of course correct.

Example 7.5 (See also [40].) Let

$$g(x, y, \varrho) = \frac{1}{2\pi\sqrt{(1-\varrho^2)}}\, e^{-\frac{1}{2(1-\varrho^2)}(x^2 - 2\varrho xy + y^2)}$$

for some ϱ satisfying $0 < \varrho < 1$. Define the joint pdf of X and Y by

$$f(x, y) = \frac{1}{2}\left(g(x, y, \varrho) + g(x, y, -\varrho)\right),$$

see Fig. 7.7 for an illustration with $\varrho = 0.8$. Then it can be shown that X and Y have standard normal distributions with $\text{Cov}(X, Y) = 0$, but X and Y cannot be independent since $f(x, y) \neq \phi(x)\phi(y)$. The joint pdf $f(x, y)$ is a further example for a distribution of (X, Y) not being bivariate normal, but admitting individual normal distributions of X and Y.   □

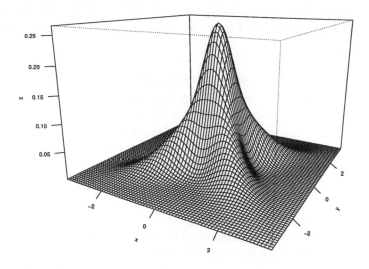

Fig. 7.7: Joint pdf of X and Y in Example 7.5

Example 7.6 (See also [49].) Let X have standard normal distribution and define the random variable y as

$$Y = \begin{cases} X & \text{if } |X| \le c \\ -X & \text{otherwise} \end{cases}$$

for some fixed number $c > 0$. Then, by symmetry Y has standard normal distribution. In addition, it can be shown that

$$\text{Cov}(X, Y) = \text{E}(XY) = 4\Phi(c) - 4c\phi(c) - 3 .$$

Therefore X and Y are uncorrelated if c satisfies the equation $\Phi(c) = 0.75 + c\phi(c)$, an approximate solution being $c = 1.538172254$. Of course, X and Y can never be independent. □

Two further examples are given in [11], see also [48].

7.2.3 Does Uncorrelatedness of Two Normals Imply Independence Only If They Are Jointly Normal?

In order to emphasize that uncorrelatedness and independence of two normals are not equivalent in general (and thus to avoid the above fallacy) it is often stated that 'uncorrelatedness of two normals implies their independence *only if* the two variables have joint normal distribution'. This statement, however, implies that joint normality is the only situation under which uncorrelatedness of two normals implies their independence, which is not the case.

Example 7.7 Consider the joint pdf of X and Y from Example 7.3. Then X and Y both have standard normal distribution. Further, it can be shown that $\mathrm{Cov}(X, Y) = \alpha/\pi$, so that X and Y are uncorrelated if and only if $\alpha = 0$. But then $f(x, y) = \phi(x)\phi(y)$, so that X and Y are necessarily independent. Hence, uncorrelatedness implies independence, although the joint pdf is not bivariate normal. □

7.2.4 Is Any Linear Combination of Two Normals Again Normal?

The answer to this question is *no, not in general*. It is correct that if X and Y have joint normal distribution, then any linear combination of X and Y is also normally distributed, but this is not the case in general.

Example 7.8 (See also [69, Sect. 2.11].) Let X have standard normal distribution. Let Y be a random variable such that $P(Y = X) = 0.5$ and $P(Y = -X) = 0.5$. Then by symmetry Y has standard normal distribution. But $P(Y = -X) = 0.5$ means $P(X + Y = 0) = 0.5$, so that the sum $X + Y$ cannot have a normal distribution. □

See also [49] for an example of having normality of exactly n (but not more) linear combinations of two normals.

7.3 Vector and Matrix Representation

A usual representation of a two-dimensional random variable (X_1, X_2) is a column vector[1] $\boldsymbol{X} = \begin{pmatrix} X_1 \\ X_2 \end{pmatrix}$. Then

$$\mu = \begin{pmatrix} \mathrm{E}(X_1) \\ \mathrm{E}(X_2) \end{pmatrix} \quad \text{and} \quad \Sigma = \begin{pmatrix} \mathrm{Var}(X_1) & \mathrm{Cov}(X_1, X_2) \\ \mathrm{Cov}(X_2, X_1) & \mathrm{Var}(X_2) \end{pmatrix}$$

are called *expectation vector* and *covariance matrix*[2] (also called *variance matrix* or *variance-covariance matrix*) of $\boldsymbol{X}$. By using

$$\Sigma = \begin{pmatrix} \sigma_1^2 & \varrho\sigma_1\sigma_2 \\ \varrho\sigma_1\sigma_2 & \sigma_2^2 \end{pmatrix}, \quad \Sigma^{-1} = \frac{1}{\det(\Sigma)} \begin{pmatrix} \sigma_2^2 & -\varrho\sigma_1\sigma_2 \\ -\varrho\sigma_1\sigma_2 & \sigma_1^2 \end{pmatrix},$$

where
$$\det(\Sigma) = \sigma_1^2\sigma_2^2(1 - \varrho^2),$$

the pdf of the bivariate normal distribution may also be written as

$$f(\boldsymbol{x}) = [\det(2\pi\Sigma)]^{-\frac{1}{2}} \, \mathrm{e}^{-\frac{1}{2}(\boldsymbol{x}-\mu)'\Sigma^{-1}(\boldsymbol{x}-\mu)}, \quad \boldsymbol{x} = \begin{pmatrix} x_1 \\ x_2 \end{pmatrix}.$$

Compared with the pdf of the normal distribution

$$f(x) = [2\pi\sigma^2]^{-\frac{1}{2}} \, \mathrm{e}^{-\frac{1}{2}(x-\mu)(\sigma^2)^{-1}(x-\mu)},$$

the former may also be obtained from the latter by an appropriate replacement of quantities by vectors and matrices.

This naturally carries over to a k- instead of a two-dimensional random variable. Namely, the k-dimensional random variable $\boldsymbol{X} = (X_1, \ldots, X_k)'$ is said to have a *k-variate normal distribution* (*multivariate normal distribution*), if it has pdf

---

[1] Then the transpose of $\boldsymbol{X}$ is a row vector $\boldsymbol{X}' = (X_1, X_2)$. This is in a slight contradiction with our former notation of a two-dimensional random variable as a row vector without introducing the transposition symbol. However, the contradiction only refers to notational convention and not to the stated facts. Nonetheless, when vector-matrix computations are applied, the notational difference between a column and a row vector is of importance.

[2] Note that $\mathrm{Cov}(X_1, X_2) = \mathrm{Cov}(X_2, X_1)$

$$f(x) = [\det(2\pi\Sigma)]^{-1/2} e^{-\frac{1}{2}(x-\mu)'\Sigma^{-1}(x-\mu)}, \quad x = \begin{pmatrix} x_1 \\ \vdots \\ x_x \end{pmatrix},$$

for some $k \times 1$ vector μ and some $k \times k$ symmetric positive definite matrix Σ. A contour of the k-variate normal distribution

$$\{x : (x - \mu)'\Sigma^{-1}(x - \mu) = c^2\}$$

is a k-dimensional ellipsoid for any fixed c. The k-variate normal distribution plays an important role in multivariate analysis, see [86, 47]. It may even be generalized to the case of singular symmetric nonnegative definite matrices Σ, in which case the pdf cannot be written down in the above form, since Σ^{-1} does not exist. See also [28, p. 360] for some remarks concerning the occurrence of singular covariance matrices in linear regression.

Transformations

When Z_1 and Z_2 are two independent random variables with standard normal distribution, then $Z = (Z_1, Z_2)'$ has a bivariate normal distribution with parameters $\mu_1 = \mu_2 = 0$, $\sigma_1 = \sigma_2 = 1$ and $\varrho = 0$. When

$$H = \begin{pmatrix} \sigma_1 & 0 \\ \sigma_2\varrho & \sigma_2\sqrt{1-\varrho^2} \end{pmatrix}, \quad \mu = \begin{pmatrix} \mu_1 \\ \mu_2 \end{pmatrix},$$

then the transformation $HZ + \mu$ has a bivariate normal distribution with expectation vector μ and covariance matrix

$$\Sigma = HH' = \begin{pmatrix} \sigma_1^2 & \varrho\sigma_1\sigma_2 \\ \varrho\sigma_1\sigma_2 & \sigma_2^2 \end{pmatrix},$$

i.e. with parameters μ_1, μ_2, σ_1, σ_2 and ϱ. This transformation is useful for generating a sample from the bivariate normal distribution, see Sect. 7.1.6. Other transformations, e.g.

$$H = \frac{1}{\sqrt{2}} \begin{pmatrix} \sigma_1\sqrt{1+\varrho} & -\sigma_1\sqrt{1-\varrho} \\ \sigma_2\sqrt{1+\varrho} & \sigma_2\sqrt{1-\varrho} \end{pmatrix}$$

are also possible.

Chapter 8

Transformations of Normal Variables

In this chapter we consider the distribution of some elementary functions of a normally distributed random variable, as well as the distribution of two and more independent normally distributed random variables.

8.1 Functions of a Random Variable

Suppose that we know the pdf $f(x)$ of a continuous random variable X, and we are interested in the pdf of a specific function $Y = g(X)$ of X. Then under certain assumptions we may obtain this pdf by the so-called *transformation technique* described as follows.

- First, we have to identify the *support* of the distribution of X, that is the set of values $\mathcal{X}$ for which the pdf is strictly greater than zero, i.e.

$$\mathcal{X} = \{x : f(x) > 0\}.$$

- Second, we consider only functions of X which are strictly monotone on $\mathcal{X}$, i.e. for $, u, v \in \mathcal{X}$ we either have $u > v \Rightarrow g(u) > g(v)$ or $u < v \Rightarrow g(u) > g(v)$. But this is a rather strong assumption, many simple functions $g(\cdot)$ usually being not strictly monotone on the whole set $\mathcal{X}$. Fortunately, it is possible to relax this assumption in such a way that the function $g(\cdot)$ is only required to be piecewise strictly monotone on subsets $A_1, \ldots, A_k$ of $\mathcal{X}$, such that the sets $A_0, A_1, \ldots, A_k$ constitute a partition of $\mathcal{X}$, where $A_0 \subset \mathcal{X}$ satisfies $P(X \in A_0) = 0$. (The introduction of A_0 offers the possibility for ignoring the strict monotonicity assumption on this set. Often A_0 consists only of one point.) We must further assume that the pdf $f(x)$ is continuous on each A_i.

- Third, we identify the support induced by the function $g(\cdot)$ with respect to A_i, being the set

$$\mathcal{Y} = \{y : y = g(x) \text{ for some } x \in A_i\} ,$$

which is required to be the same for each $i = 1, \ldots, k$.

- Fourth, if the induced support $\mathcal{Y}$ is the same with respect to each A_i, then we compute the inverse functions $x = g_i^{-1}(y)$ of $g(\cdot)$ for each set set A_i, $i = 1, \ldots, k$. If every $g_i^{-1}(y)$ has a continuous derivative on $\mathcal{Y}$, then the pdf of $Y = g(X)$ is given as

$$f(y) = \begin{cases} \sum_{i=1}^{k} |\frac{\mathrm{d}}{\mathrm{d}y} g_i^{-1}(y)| f(g_i^{-1}(y)) & y \in \mathcal{Y} \\ 0 & \text{otherwise} \end{cases} .$$

The above is a slightly less formal description of the transformation technique, as for example given in [12, Sect. 2.1]. It is also possible to generalize this technique to functions of more than one random variable, see e.g. [12, Chapt. 4].

Example 8.1 Suppose that a continuous random variable X has continuous pdf $f(x)$ with support $\mathcal{X}$ being the interval $(-\infty, \infty)$. What is the pdf of $Y = g(X) = 1/X^2$?

We may use the partition $(-\infty, \infty) = A_0 \cup A_1 \cup A_2$ with $A_0 = \{0\}$, $A_1 = (-\infty, 0)$ and $A_2 = (0, \infty)$. Then the function $g(\cdot)$ is strictly increasing on A_1 and strictly decreasing on A_2. Since X is continuous, $P(X \in A_0) = P(X = 0) = 0$. The induced support with respect to $g(\cdot)$ and A_1 is

$$\mathcal{Y} = \{y : y = 1/x^2 \text{ for some } x \in A_1\} = (0, \infty) ,$$

being the same as the induced support with respect to $g(\cdot)$ and A_2. The inverse functions of $g(\cdot)$ with respect to A_1 and A_2 are $g_1^{-1}(y) = -1/\sqrt{y}$ and $g_2^{-1}(y) = 1/\sqrt{y}$, respectively. Then the pdf of Y is given as

$$f(y) = \left(\left| \frac{1}{2\sqrt{y^3}} \right| f(-1/\sqrt{y}) + \left| -\frac{1}{2\sqrt{y^3}} \right| f(1/\sqrt{y}) \right) I_{(0,\infty)}(y)$$

$$= \frac{1}{2\sqrt{y^3}} \left(f(-1/\sqrt{y}) + f(1/\sqrt{y}) \right) I_{(0,\infty)}(y) \ .$$

If, for example, the pdf of X is $f(x) = \frac{1}{2} e^{-|x|}$, then the pdf of $Y = 1/X^2$ is $f(y) = \frac{1}{2\sqrt{y^3}} e^{-1/\sqrt{y}} I_{(0,\infty)}(y)$. $\quad\square$

8.2 Functions of a Single Normal Variable

For a normally distributed random variable X, the distributions of $|X|$, $1/X$ and X^2 are discussed.

8.2.1 Distribution of the Absolute Value

Suppose X has a $N(\mu, \sigma^2)$ distribution. What is the distribution of $|X|$? By an application of the transformation technique, the pdf of the distribution of $|X|$ can be written as

$$f(x) = \frac{1}{\sqrt{2\pi}\sigma} \left(e^{-\frac{1}{2}(x+\mu)^2/\sigma^2} + e^{-\frac{1}{2}(x-\mu)^2/\sigma^2} \right) I_{[0,\infty)}(x) \ .$$

Since the pdf is the same irrespective of the sign of μ, we may also write

$$f(x) = \frac{1}{\sigma} \left[\phi\left(\frac{x+|\mu|}{\sigma} \right) + \phi\left(\frac{x-|\mu|}{\sigma} \right) \right] I_{[0,\infty)}(x) \ .$$

When $|\mu| > 3\sigma$, then the first summand in $f(x)$ is quite small, so that $f(x)$ is nearly the pdf of the $N(|\mu|, \sigma^2)$ distribution, see also Fig. 8.1.

As noted in [42], the distribution of the absolute value $|X|$ can be of interest, when X is a variable being assumed to follow a normal distribution, but for some reason the observations are recorded without their algebraic sign[1]. Leone et al. [42] give examples for such situations and call the distribution of $|X|$ the *folded normal distribution*.

---

[1] So that in fact absolute values are recorded.

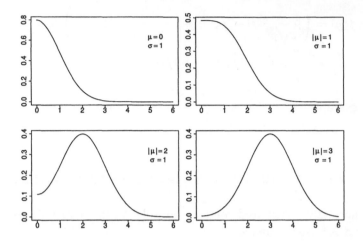

Fig. 8.1: Probability density functions of the absolute value of a $N(\mu, \sigma^2)$ distributed random variable

Expectation and Variance

The expectation of $|X|$ is

$$
\begin{aligned}
\mathrm{E}(|X|) &= \int_0^\infty x \frac{1}{\sigma} \left(\phi\left(\frac{x + |\mu|}{\sigma} \right) + \phi\left(\frac{x - |\mu|}{\sigma} \right) \right) \mathrm{d}x \\
&= \frac{1}{\sigma} \left[\int_0^\infty x\phi\left(\frac{x + |\mu|}{\sigma} \right) \mathrm{d}x + \int_0^\infty x\phi\left(\frac{x - |\mu|}{\sigma} \right) \mathrm{d}x \right]
\end{aligned}
$$

By applying the indefinite normal integral

$$
\int x\phi(a + bx)\mathrm{d}x = -\frac{1}{b^2}\phi(a + bx) - \frac{a}{b^2}\Phi(a + bx) ,
$$

straightforward calculations yield

$$
\mathrm{E}(|X|) = 2\sigma\phi(|\mu|/\sigma) + |\mu|(2\Phi(|\mu|/\sigma) - 1) ,
$$

which may also be written as

$$
\mathrm{E}(|X|) = 2\sigma\phi(\mu/\sigma) + \mu(1 - 2\Phi(-\mu/\sigma)) .
$$

The expectation of $|X|^2$ is the same as that of X^2, being $E(|X|^2) = \mu^2 + \sigma^2$. Then a formula for the variance of $|X|$ can be obtained from

$$\mathrm{Var}(|X|) = \mu^2 + \sigma^2 - E(|X|)^2 .$$

Again, when $|\mu| > 3\sigma$, then $\phi(|\mu|/\sigma) \approx 0$ and $\Phi(|\mu|/\sigma) \approx 1$, so that $E(|X|) \approx |\mu|$, corresponding to the above conclusion that for $|\mu| > 3\sigma$ the distribution of $|X|$ is nearly $N(|\mu|, \sigma^2)$.

Half-Normal Distribution

When $\mu = 0$, compare also Fig. 8.1, then the pdf of $|X|$ is

$$f(x) = \frac{2}{\sqrt{2\pi}\sigma} e^{-\frac{1}{2}x^2/\sigma^2} I_{[0,\infty)}(x) .$$

Since $f(x)$ may be written as $f(x) = (1/\sigma)\eta(x/\sigma)$, where

$$\eta(x) = \frac{2}{\sqrt{2\pi}} e^{-\frac{1}{2}x^2} I_{[0,\infty)}(x) ,$$

the set $\{f(x) : 0 < \sigma^2\}$ is a scale family with standard pdf $\eta(x)$. The pdf $\eta(x)$ is also the pdf of $|Z|$ when Z has a standard normal distribution. It may be made the standard pdf of a location-scale family of pdfs $f(x)$ with $f(x) = (1/\sigma)\eta[(x - \alpha)/\sigma]$, so that

$$f(x; \alpha, \sigma) = \frac{2}{\sqrt{2\pi}\sigma} e^{-\frac{1}{2}(x-\alpha)^2/\sigma^2} I_{[\alpha,\infty)}(x), \quad -\infty < \alpha < \infty, 0 < \sigma .$$

This is the pdf of a *half-normal distribution* with location parameter α and scale parameter σ.

Theorem 8.1 *Let Z have a standard normal distribution. Then $\alpha + \sigma|Z|$ has a half-normal distribution with location parameter α and scale parameter σ.*

8.2.2 Distribution of the Square

Suppose X has a $N(\mu, \sigma^2)$ distribution. Then the application of the transformation technique shows that X^2 has pdf

$$f(x) = \frac{1}{2\sigma\sqrt{x}}\left[\phi\left(\frac{\sqrt{x}+\mu}{\sigma}\right) + \phi\left(\frac{\sqrt{x}-\mu}{\sigma}\right)\right]I_{(0,\infty)}(x) .$$

For $\mu = 0$ and $\sigma = 1$, the pdf can be written as

$$f(x) = \frac{1}{\sqrt{\pi}\, 2^{1/2}} x^{-1/2}\, e^{-x/2} I_{(0,\infty)}(x) .$$

By using $\Gamma(1/2) = \sqrt{\pi}$ it may also be expressed as

$$f(x) = \frac{1}{\Gamma(\nu/2)\, 2^{\nu/2}} x^{\nu/2-1}\, e^{-x/2} I_{(0,\infty)}(x)$$

for $\nu = 1$. For a positive integer ν, this is the pdf of the *chi-square distribution* with parameter ν.

Theorem 8.2 *Let X have a standard normal distribution. Then X^2 has a chi-square distribution with parameter $\nu = 1$.*

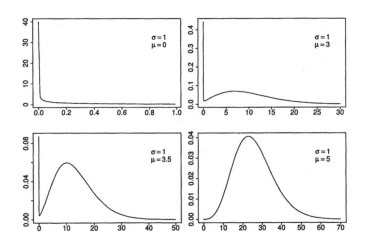

Fig. 8.2: Probability density functions of the square of a $N(\mu, \sigma^2)$ distributed random variable

Expectation and Variance

When X has a normal distribution with parameters μ and σ^2, then from Sect. 2.3.2 we obtain $E(X^2) = \mu^2 + \sigma^2$ and $E(X^4) = 3\sigma^4 + 6\mu^2\sigma^2 + \mu^4$. Hence expectation and variance of X^2 are given as

$$E(X^2) = \mu^2 + \sigma^2 \quad \text{and} \quad \text{Var}(X^2) = 2\sigma^2(2\mu^2 + \sigma^2) \ .$$

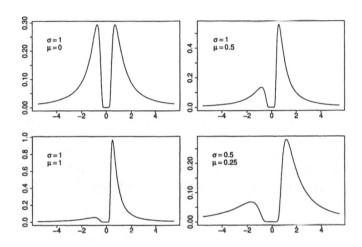

Fig. 8.3: Probability density functions of the reciprocal of a $N(\mu, \sigma^2)$ distributed random variable

8.2.3 Distribution of the Reciprocal

Suppose X has a $N(\mu, \sigma^2)$ distribution. Then the application of the transformation technique shows that $1/X$ has pdf

$$f(x) = \frac{1}{\sqrt{2\pi}\sigma x^2} e^{-\frac{1}{2}((1/x)-\mu)^2/\sigma^2}, \quad x \neq 0 \ .$$

As also noted by Robert [68] who considers a more general class of pdfs, the distribution is always bi-modal with modes at

$$z_{1,2} = \frac{\pm\sqrt{\mu^2 + 8\sigma^2} - \mu}{4\sigma^2} ,$$

see Fig. 8.3. Expectation and variance do not exist.

The distribution of the reciprocal of a normal variable is not to be confounded with the so-called *inverse Gaussian* distribution, having pdf

$$f(x) = \left[\frac{\lambda}{2\pi x^3}\right]^{1/2} \exp\left[-\frac{\lambda}{2\mu^2 x}(x-\mu)^2\right] I_{(0,\infty)}(x), \quad \mu > 0, \lambda > 0 .$$

The name 'inverse Gaussian' refers to an inverse relationship between cumulant generating functions of the inverse Gaussian and a specific normal distribution, see e.g. [37, Chapter 15].

8.2.4 Distribution of Minimum and Maximum

If $X_1, \ldots, X_n$ is a sample from the normal distribution with parameters μ and σ^2, then the minimum of the sample is the first order statistic, while the maximum is the n-th order statistic. Then from Sect. 2.4.2, the pdf of the minimum is

$$f_{(1)}(x) = \frac{n}{\sigma}\phi\left(\frac{x-\mu}{\sigma}\right)\left[1 - \Phi\left(\frac{x-\mu}{\sigma}\right)\right]^{n-1} ,$$

while the pdf of the maximum is

$$f_{(n)}(x) = \frac{n}{\sigma}\phi\left(\frac{x-\mu}{\sigma}\right)\left[\Phi\left(\frac{x-\mu}{\sigma}\right)\right]^{n-1} .$$

The moments can only be computed numerically. The pdf of the minimum is skewed to the left, while the pdf of the maximum is skewed to the right.

For illustration purposes, let us consider 1000 samples of size $n = 100$ from the standard normal distribution. For each of the samples the minimum and the maximum is computed. The respective values are shown in Figures 8.4 (a) and (b). Using the numerical integration procedures from MAPLE 6, the expectations of the minimum and the maximum are -2.5076 and 2.5076, while the standard deviation of minimum as well as maximum is 0.429424.

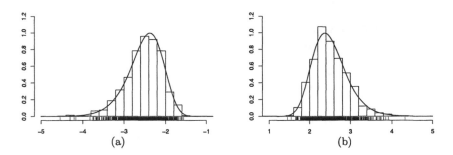

Fig. 8.4: Histogram of minima (a) and maxima (b) of 1000 samples from the standard normal distribution of size $n = 100$, superimposed by the respective probability density function

8.3 Functions of Two Independent Normals

Let X and Y be jointly distributed continuous random variables with pdf $f(x, y)$. Then, it can be shown that the pdfs of $U = X + Y$, $V = XY$ and $W = X/Y$, are given as

$$f(u) = \int_{-\infty}^{\infty} f(x, u - x)\, \mathrm{d}x = \int_{-\infty}^{\infty} f(u - y, y)\, \mathrm{d}y\ ,$$

$$f(v) = \int_{-\infty}^{\infty} \frac{1}{|x|} f(x, \frac{v}{x})\mathrm{d}x = \int_{-\infty}^{\infty} \frac{1}{|y|} f(\frac{v}{y}, y)\mathrm{d}y\ ,$$

and

$$f(w) = \int_{-\infty}^{\infty} |y| f(wy, y)\mathrm{d}y\ ,$$

see e.g. [54, Chapter V]. When X and Y are independent of each other, the joint pdf is the product of the individual pdfs, and hence if X and Y are independently $N(\mu_1, \sigma_1^2)$ and $N(\mu_2, \sigma_2^2)$ distributed, the joint pdf of X and Y is

$$f(x, y) = \frac{1}{\sigma_1 \sigma_2} \phi\left(\frac{x - \mu_1}{\sigma_1}\right) \phi\left(\frac{y - \mu_2}{\sigma_2}\right)\ .$$

In the subsequent section we shortly discuss the distribution of U, V and W, provided that X and Y are independent.

8.3.1 Distribution of the Sum

As noted above, the pdf of the sum $U = X + Y$ of two independent normal variables X and Y is given as

$$f(u) = \int_{-\infty}^{\infty} f(x, u - x) \, dx = \frac{1}{\sigma_1 \sigma_2} \int_{-\infty}^{\infty} \phi\left(\frac{x - \mu_1}{\sigma_1}\right) \phi\left(\frac{u - x - \mu_2}{\sigma_2}\right) dx \,,$$

which may also be written as

$$f(u) = \frac{1}{\sigma_2} \int_{-\infty}^{\infty} \phi(t) \, \phi\left(\frac{u - (\mu_1 + \mu_2)}{\sigma_2} - \frac{\sigma_1}{\sigma_2} t\right) dt.$$

By applying the indefinite normal integral

$$\int \phi(x) \, \phi(a + bx) \, dx = c^{-1} \phi(a/c) \Phi(cx + a/c), \quad c = \sqrt{1 + b^2} \,,$$

straightforward calculations yield

$$f(u) = \frac{1}{\sqrt{\sigma_1^2 + \sigma_2^2}} \, \phi\left(\frac{u - (\mu_1 + \mu_2)}{\sqrt{\sigma_1^2 + \sigma_2^2}}\right) .$$

Hence, $U = X + Y$ is also normally distributed with $\mathrm{E}(U) = \mu_1 + \mu_2$ and $\mathrm{Var}(U) = \sigma_1^2 + \sigma_2^2$.

Theorem 8.3 *Let X be $N(\mu_1, \sigma_1^2)$ and Y be $N(\mu_2, \sigma^2)$ distributed, and let X and Y be independent. The $X + Y$ is $N(\mu_1 + \mu_2, \sigma_1^2 + \sigma_2^2)$ distributed.*

Note that the above theorem is nothing else but a special case of Theorem 7.2.

8.3.2 Distribution of Product and Quotient

For simplicity we consider the distribution of product and quotient only for the case that X and Y have a standard normal distribution and are independent of each other. There exist more general results on the pdf of product and quotient even for dependent normally distributed random variables when the joint pdf is *bivariate normal.*

Distribution of the Product

It can be shown that if X and Y are two independent random variables having standard normal distribution, then the pdf of the product $V = XY$ is

$$f(v) = \frac{1}{\pi} K_0(|v|), \quad v \neq 0,\ ,$$

where $K_0(v)$, $v > 0$, is the modified Bessel function[2] of the second kind to the order 0. This function is implemented in many mathematical and statistical software packages (often named as 'BesselK'). Since X and Y are independent,

$$E(V) = 0 \quad \text{and} \quad \text{Var}(V) = 1\,,$$

see e.g. [54, Sect. V 2.3]. Fig. 8.5 shows a histogram of $n = 1000$ generated values from $f(v)$ together with the pdf.

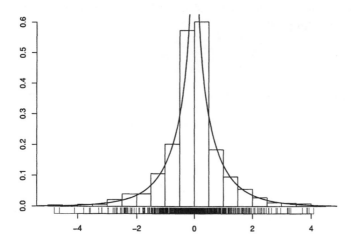

Fig. 8.5: Histogram of a sample with size $n = 1000$ from the product of two independent standard normal variables together with the corresponding pdf

---

[2]One possible integral representation for $v > 0$ is $K_0(v) = \int_0^\infty \frac{\cos(vt)}{\sqrt{t^2+1}}\, dt$.

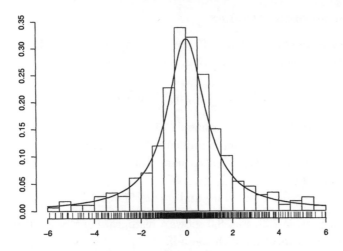

Fig. 8.6: Histogram of 890 values out of a sample of size $n = 1000$ from the quotient of two independent standard normal variables together with the corresponding pdf

Distribution of the Quotient

When X and Y are two independent random variable having standard normal distribution, then the pdf of the quotient $W = X/Y$ has pdf

$$f(w) = \frac{1}{\pi} \frac{1}{w^2 + 1} \, .$$

This is the standard pdf with respect to the location-scale family of pdfs given as

$$h(w) = \frac{1}{\pi b} \frac{1}{(w - a)^2 / b^2 + 1}$$

with location parameter $-\infty < a < \infty$ and scale parameter $0 < b$. A random variable having pdf $h(w)$ is said to be *Cauchy distributed*. Moreover, by using $\Gamma(1) = 1$ and $\Gamma(1/2) = \sqrt{\pi}$, the pdf $f(w)$ may be written as

$$f(w) = \frac{\Gamma\left(\frac{\nu+1}{2}\right)}{\Gamma\left(\frac{\nu}{2}\right)} \frac{1}{\sqrt{\nu\pi}} \frac{1}{(w^2/\nu)^{(\nu+1)/2}}$$

for $\nu = 1$. For an arbitrary positive integer ν, this is the pdf of the t *distribution* with parameter ν.

It is well known that the expectation and variance of the Cauchy distribution are not finite. Nonetheless, the parameter a is the center of the distribution and from the cumulative distribution function

$$F(w) = \frac{1}{2} + \frac{1}{\pi} \arctan\left(\frac{w-a}{b}\right)$$

it is easy to compute probabilities $F(a+k) - F(a-k)$ for a Cauchy distributed random variable to realize within an interval $[a-k, a+k]$ for any k. For example for the standard Cauchy distribution ($a = 0$ and $b = 1$), we have a probability of 0.8949 for an outcome within the interval $[-6, 6]$. This is reflected by sample represented in Fig. 8.6, where 890 out of 1000 values lie in $[-6, 6]$. Since the Cauchy distribution has heavy tails, it is not unlikely that in a given sample some observations are far away from the center a. The 10 smallest and 10 greatest values of the sample corresponding to Fig. 8.6 are given in Table 8.1.

-546.74	-328.27	-271.75	-181.97	-121.27	-75.25	-58.96	-51.62	-42.55	-36.44
...									...
31.08	35.62	35.92	36.16	36.27	38.52	48.98	49.76	121.40	180.01

Table 8.1: The 10 smallest and 10 greatest values of the sample represented in Fig. 8.6

8.4 Functions Related to a Normal Sample

In this sections we consider the distribution of functions of a random sample $X_1, \ldots, X_n$ from the normal distribution.

8.4.1 Standard Normal Sample

The *chi-square distribution* is the distribution of the sum of squares $\sum_{i=1}^{n} X_i^2$ when $X_1, \ldots, X_n$ is a sample from the standard normal distribution. *Student's t* and *Snedecor's F distribution* may be derived in relationship to the chi-square distribution.

Chi-Square Distribution

We have already noted that the square of a random variable with standard normal distribution has a chi-square distribution with parameter $\nu = 1$. The chi-square distribution with arbitrary positive integer ν is the distribution of the sum of squares of a random sample of size ν from the standard normal distribution.

Definition 8.1 *Let* $X_1, \ldots, X_n$ *be a sample from the standard normal distribution. Then the distribution of*

$$Z = \sum_{i=1}^{n} X_i^2$$

is called the chi-square distribution[3] *with parameter* $\nu = n$ *(denoted by* $\chi^2_{(\nu)}$*).*

The pdf of the $\chi^2_{(\nu)}$ distribution is given as

$$f(x) = \frac{1}{\Gamma(\nu/2)\, 2^{\nu/2}}\, x^{\nu/2-1}\, e^{-x/2} I_{(0,\infty)}(x) \ .$$

It has expectation ν and variance 2ν. The parameter ν is usually referred as the 'degrees of freedom'. In general, when the chi-square distribution with parameter ν is addressed the degrees of freedom ν may also be seen as the number of independent squares[4] involved in the sum of random variables.

---

[3]The chi-square distribution is also called *Helmert distribution* in view of the contributions by Friedrich Robert Helmert (1843–1917). See also [36, Sect. 18.1] for a short historical review of the chi-square distribution.

[4]This is not necessarily identical to the sample size n. For example, if $X_1, \ldots, X_n$ is a sample from the standard normal distribution, then it can be shown that

$$Z = \sum_{i=1}^{n} (X_i - \overline{X})^2$$

has a chi-square distribution with $\nu = n - 1$ degrees of freedom, and that the sum involves $n - 1$ and not n independent squares, since for any i it is possible to compute $X_i - \overline{X}$ from the other $X_1 - \overline{X}, \ldots X_{i-1} - \overline{X}, X_{i+1} - \overline{X}, \ldots, X_n - \overline{X}$, see also [54, Sect. VI 4.3] for a discussion.

Cumulative Distribution Function

The cdf of the chi-square distribution for $x > 0$ is given by

$$F(x) = \frac{\gamma(\nu/2, x/2)}{\Gamma(\nu/2)},$$

where

$$\gamma(\alpha, x) = \int_0^x t^{\alpha-1} e^{-t} dt$$

is the *lower incomplete gamma function*.

Normal Approximations

The pdf of the chi-square distribution can also be written as

$$f(x) = \frac{\lambda^n}{\Gamma(n)} x^{n-1} e^{-\lambda x} I_{(0,\infty)}(x)$$

for $\lambda = 1/2$ and $n = \nu/2$. It is therefore a special case of the gamma distribution. As already noted in Example 2.2, p. 73, for larger values of ν it can thus be approximated by the $N(\nu, 2\nu)$ distribution. This is, however, known to be not very good unless ν is rather large. Specific functions of a $\chi^2_{(\nu)}$ distributed random variable X admit more accurate normal approximations, the most common being $\sqrt{2X}$ having an approximate $N(\sqrt{2\nu - 1}, 1)$ distribution according to Fisher [22], and $(X/\nu)^{1/3}$ having an approximate $N(1 - \frac{2}{9\nu}, \frac{2}{9\nu})$ distribution according to Wilson and Hilferty [91], the latter being a better approximation than the former. The cdf of the chi-square distribution may thus also be approximated by

$$F(x) \approx \Phi\left(\sqrt{\frac{9\nu}{2}}\left(\left(\frac{x}{\nu}\right)^{1/3} - 1 + \frac{2}{9\nu}\right)\right),$$

from which one may also obtain approximations for p-quantiles of the chi-square distribution.

Chi Distribution

When X has a chi-square distribution with parameter ν, the distribution of the positive square root $\sqrt{X}$ is called the *chi-distribution* with parameter

ν and is denoted by $\chi_{(\nu)}$. By applying the transformation technique, its pdf is shown to be

$$f(x) = \frac{1}{\Gamma(\nu/2)\, 2^{(\nu/2)-1}}\, x^{\nu-1}\, e^{-x^2/2} I_{(0,\infty)}(x) \,,$$

Expectation and variance of $\sqrt{X}$ having $\chi_{(\nu)}$ distribution are

$$\mathrm{E}(\sqrt{X}) = \sqrt{2}\,\frac{\Gamma\left(\frac{\nu+1}{2}\right)}{\Gamma\left(\frac{\nu}{2}\right)} \quad \text{and} \quad \mathrm{Var}(\sqrt{X}) = \nu - 2\left(\frac{\Gamma\left(\frac{\nu+1}{2}\right)}{\Gamma\left(\frac{\nu}{2}\right)}\right)^2.$$

For $\nu = 1, 2, 3$ the chi distribution is a special half-normal, Rayleigh, and Maxwell distribution, respectively.

Adjusted Chi-Square Distribution

Quite often, the random variable X/ν is regarded when X has a $\chi^2_{(\nu)}$ distribution. Applying the transformation technique, the pdf of X/ν is

$$f(x) = \frac{(\nu/2)^{\nu/2}}{\Gamma(\nu/2)}\, x^{\nu/2-1}\, e^{-(\nu/2)x} I_{(0,\infty)}(x) \,.$$

The distribution may be called *adjusted chi-square distribution*, where the adjustment is with respect to the degrees of freedom. The distribution has expectation 1 and variance $2/\nu$, vanishing for $\nu \to \infty$.

Square Root of Adjusted Chi-Square

When we consider the positive square root $\sqrt{X/\nu}$, the corresponding pdf is

$$f(x) = 2\,\frac{(\nu/2)^{\nu/2}}{\Gamma(\nu/2)}\, x^{\nu-1}\, e^{-(\nu/2)x^2} I_{(0,\infty)}(x) \,.$$

For the special case $\nu = 1$, this is the pdf of a half-normal distribution with location parameter 0 and scale parameter 1. The distribution has expectation $\gamma_\nu := \sqrt{\frac{2}{\nu}}\Gamma\left(\frac{\nu+1}{2}\right)/\Gamma\left(\frac{\nu}{2}\right)$ and variance $1 - \gamma_\nu^2$. For $\nu \to \infty$ the quantity γ_ν approaches 1.

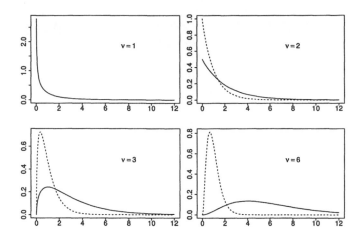

Fig. 8.7: Probability density functions of chi-square (solid line) and adjusted chi-square (dotted line) distribution with ν degrees of freedom

Student's t Distribution

We may now consider the distribution of the quotient of a standard normal variable and the positive square root of an adjusted chi-square variable when both are independent.

Definition 8.2 *Let X have standard normal distribution, let Y have $\chi^2_{(\nu)}$ distribution, and let X and Y be independent. Then the distribution of*

$$Z = \frac{X}{\sqrt{Y/\nu}}$$

is called Student's t distribution[5] *with parameter ν (denoted by $t_{(\nu)}$).*

---

[5]The t distribution has been developed by William Sealey Gosset (1876–1937) who published his result 1908 under the pseudonym 'Student' [84]. See also the paragraph following our Theorem 8.5. The term 'Student's distribution' or 'Student's t distribution' has been popularized by Ronald Aylmer Fisher (1890–1962), who probably used the letter 't' for no other reason but introducing a symbol different from his already employed letter 'z'.

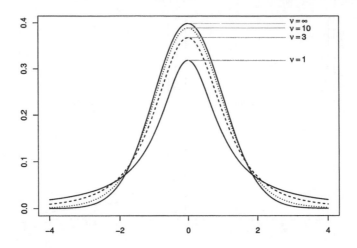

Fig. 8.8: Probability density functions of Student's t distribution with ν degrees of freedom

Usually the parameter ν is also called 'degrees of freedom'. The pdf of the $t_{(\nu)}$ distribution is given as

$$f(x) = \frac{\Gamma\left(\frac{\nu+1}{2}\right)}{\Gamma\left(\frac{\nu}{2}\right)} \frac{1}{\sqrt{\nu\pi}} \frac{1}{(1 + x^2/\nu)^{(\nu+1)/2}}.$$

It has expectation 0 for $\nu > 1$ and variance $\nu/(\nu - 2)$ for $\mu > 2$. For the choice $\nu = 1$, the $t_{(\nu)}$ distribution is the Cauchy distribution (with location 0 and scale 1). For $\nu \to \infty$ the t distribution approaches the standard normal distribution.

Snedecor's F Distribution

Eventually we consider the distribution of two independent adjusted chi-square variables.

Definition 8.3 *Let X have $\chi^2_{(\lambda)}$ distribution, let Y have $\chi^2_{(\nu)}$ distribution, and let X and Y be independent. Then the distribution of*

$$Z = \frac{X/\lambda}{Y/\nu}$$

is called Snedecor's F distribution[6] *with parameters* λ *and* ν *(denoted by* $F_{(\lambda,\nu)}$*).*

The pdf of the $F_{(\lambda,\nu)}$ distribution is given as

$$f(x) = \frac{\Gamma\left(\frac{\lambda+\nu}{2}\right)}{\Gamma\left(\frac{\lambda}{2}\right)\Gamma\left(\frac{\nu}{2}\right)} \left(\frac{\lambda}{\nu}\right)^{\lambda/2} \frac{x^{(\lambda-2)/2}}{(\frac{\lambda}{\nu}x+1)^{(\lambda+\nu)/2}} I_{(0,\infty)}(x) \, .$$

It has expectation $\nu/(\nu-2)$ for $\nu > 2$ and variance $\frac{2\nu^2(\lambda+\nu-2)}{\lambda(\nu-2)^2(\nu-4)}$ for $\nu > 4$.

Remark 8.1 For fixed positive integers λ and ν, the $F_{(\lambda,\nu)}$ and the $F_{(\nu,\lambda)}$ distributions are not the same, so that the order in which the two parameters are given is of importance. As a matter of fact, if X has $F_{(\lambda,\nu)}$ distribution, the reciprocal $1/X$ has $F_{(\nu,\lambda)}$ distribution.

Remark 8.2 If X has $t_{(\nu)}$ distribution, then X^2 has $F_{(1,\nu)}$ distribution.

Remark 8.3 If X has $F_{(\lambda,\nu)}$ distribution, then the distribution of λX approaches the $\chi^2_{(\lambda)}$ distribution for $\nu \to \infty$.

The Quotient of Independent Chi-squares

Snedecor's F distribution is the distribution of the quotient of two independent random variables having adjusted chi-square distributions. When we are interested in the distribution of the quotient of two independent random variables X and Y having (non-adjusted) chi-square distributions, then this may easily be derived from the F distribution. If X and Y are given as in Definition 8.3, then the quotient

$$U = \frac{X}{Y}$$

---

[6]The F distribution has been tabulated in 1934 by George Waddell Snedecor (1882–1974), see [78], who used the letter 'F' in honor of R.A. Fisher. Consequently, the F distribution is sometimes also called 'Fisher's distribution'. This is not to be confounded with the related 'Fisher's z distribution', see p. 204.

is equal to $(\lambda/\nu)Z$, where Z is defined in Definition 8.3. Hence the pdf of U is given as $(\nu/\lambda)f(x\nu/\lambda)$ when $f(\cdot)$ denotes the pdf of the $F_{(\lambda,\nu)}$ distribution. Therefore, the pdf of the quotient $U = X/Y$ is given as

$$f(x) = \frac{\Gamma\left(\frac{\lambda+\nu}{2}\right)}{\Gamma\left(\frac{\lambda}{2}\right)\Gamma\left(\frac{\nu}{2}\right)}\frac{x^{(\lambda-2)/2}}{(x+1)^{(\lambda+\nu)/2}}I_{(0,\infty)}(x) .$$

The pdf of the transformation

$$\frac{X}{X+Y} = \frac{U}{1+U} = \frac{\frac{\lambda}{\nu}Z}{1+\frac{\lambda}{\nu}Z}$$

is given as

$$f(x) = \frac{\Gamma\left(\frac{\lambda+\nu}{2}\right)}{\Gamma\left(\frac{\lambda}{2}\right)\Gamma\left(\frac{\nu}{2}\right)}x^{(\lambda-2)/2}(1-x)^{(\nu-2)/2}I_{(0,1)}(x) .$$

This is known to be the pdf of the *beta distribution* with parameters $\lambda/2$ and $\nu/2$. In general, the pdf of the beta distribution with parameters p and q is

$$f(x) = \frac{1}{B(p,q)}x^{p-1}(1-x)^{q-1}I_{(0,1)}(x) ,$$

where $p > 0$, $q > 0$, and $B(\cdot,\cdot)$ is the *beta function*, i.e. $B(p,q) = \frac{\Gamma(p)\Gamma(q)}{\Gamma(p+q)}$.

Fisher's z Distribution

A random variable X taking values in $(-\infty,\infty)$ is said to have *Fisher's z distribution*[7] with parameters λ and ν if the transformation e^{2X} has Snedecor's $F_{(\lambda,\nu)}$ distribution. The pdf of Fisher's z distribution is given as

$$f(x) = \frac{\Gamma\left(\frac{\lambda+\nu}{2}\right)}{\Gamma\left(\frac{\lambda}{2}\right)\Gamma\left(\frac{\nu}{2}\right)}\frac{2\lambda^{\lambda/2}\nu^{\nu/2}e^{\lambda x}}{(\nu+\lambda e^{2x})^{(\lambda+\nu)/2}} .$$

8.4.2 Distribution of Parameter-Sample Statistic Functions

The chi-square, Student t, and Snedecor F distribution play an important role in inference concerning one or more samples from the normal distribution.

---

[7]The distribution has been introduced by R.A. Fisher in 1924, [23].

A Single Sample

Suppose now that $X_1, \ldots, X_n$ is a sample from the normal distribution with parameters μ and σ^2. Then, one may consider quantities, being functions of both, the sample $X_1, \ldots, X_n$ and the parameters μ and/or σ^2. When the distribution of such quantities is known, inference about the parameters is possible.

Theorem 8.4 *Let $X_1, \ldots, X_n$ be a sample from the normal distribution with parameters μ and σ^2. Then:*

(i) *$\overline{X}$ has a normal distribution with expectation μ and variance σ^2/n.*

(ii) *$(n-1)S^2/\sigma^2$ has a chi-square distribution with $\nu = n - 1$ degrees of freedom.*

(iii) *$\overline{X}$ and S^2 are independent.*

For several purposes it is easier to consider the distribution of the function $(n-1)S^2/\sigma^2$ instead of the distribution of S^2 itself. It is of course easy to see that the pdf of the statistic S^2 is given as

$$f(x) = \left(\frac{n-1}{2\sigma^2}\right)^{(n-1)/2} \frac{1}{\Gamma\left(\frac{n-1}{2}\right)} x^{(n-3)/2} e^{-\frac{1}{2}(n-1)x/\sigma^2} I_{(0,\infty)}(x) ,$$

naturally depending on the parameter σ^2.

From the above theorem, the function $T_1 = \sqrt{n}\,(\overline{X} - \mu)/\sigma$ has a standard normal distribution, and the function $T_2 = S^2/\sigma^2$ has an adjusted chi-square distribution with parameter $\nu = n - 1$, and T_1 and T_2 are independent. Hence, $T_1/\sqrt{T_2}$ has a $t_{(n-1)}$ distribution.

Theorem 8.5 *Let $X_1, \ldots, X_n$ be a sample from the normal distribution with parameters μ and σ^2. Then the quantity*

$$T = \sqrt{n}\,\frac{\overline{X} - \mu}{S}$$

is $t_{(n-1)}$ distributed.

Gosset in his 1908 paper [84, Sect. III] did not consider the pdf of T, but of

$$\frac{\overline{X} - \mu}{D} = \frac{1}{\sqrt{n-1}}T ,$$

where $D^2 = \frac{1}{n}\sum_{i=1}^{n}(X_i - \overline{X})^2$. This can easily be shown to be

$$f(x) = \frac{\Gamma\left(\frac{n}{2}\right)}{\Gamma\left(\frac{n-1}{2}\right)}\frac{1}{\sqrt{\pi}}\left(1 + x^2\right)^{-n/2}.$$

Two Independent Samples

As already noted above, the quantity S^2/σ^2 has the adjusted chi-square distribution with parameter $\nu = n-1$. Now suppose that we have two independent random samples $X_1, \ldots, X_{n_1}$ and $Y_1, \ldots, Y_{n_2}$ from an $N(\mu_1, \sigma_1^2)$ and an $N(\mu_2, \sigma_2^2)$ distribution, respectively. Then we may denote the respective sample variances by S_1^2 and S_2^2, being independent of each other. Since S_1^2/σ_1^2 and S_2^2/σ_2^2 have adjusted chi-square distributions with parameters $n_1 - 1$ and $n_2 - 1$, the ratio $(S_1^2/\sigma_1^2)/(S_2^2/\sigma_2^2)$ is $F_{(n_1-1, n_2-1)}$ distributed.

Theorem 8.6 *Let $X_1, \ldots, X_{n_1}$ and $Y_1, \ldots, Y_{n_2}$ be two independent random samples from an $N(\mu_1, \sigma_1^2)$ and an $N(\mu_2, \sigma_2^2)$ distribution, respectively, and let S_1^2 and S_2^2 denote the respective sample variances. Then the quantity*

$$F = \frac{S_1^2/\sigma_1^2}{S_2^2/\sigma_2^2}$$

is $F_{(n_1-1, n_2-1)}$ distributed.

When we are interested in a comparison of the expectations μ_1 and μ_2, and in addition have $\sigma_1^2 = \sigma_2^2$, then inference may be based on a standardized version of the difference $\overline{X} - \overline{Y}$, where $\overline{X}$ and $\overline{Y}$ are the respective sample means.

Theorem 8.7 *Let $X_1, \ldots, X_{n_1}$ and $Y_1, \ldots, Y_{n_2}$ be two independent random samples from an $N(\mu_1, \sigma_1^2)$ and an $N(\mu_2, \sigma_2^2)$ distribution, respectively, and let $\overline{X}$ and $\overline{Y}$ denote the respective sample means. If $\mu_1 = \mu_2$ and $\sigma_1^2 = \sigma_2^2$, then the quantity*

$$T = \sqrt{\frac{n_1 n_2}{(n_1 + n_2)}}\;\frac{\overline{X}_1 - \overline{X}_2}{\frac{1}{\sqrt{n_1+n_2-2}}\sqrt{\sum_{i=1}^{n_1}(X_i - \overline{X})^2 + \sum_{i=1}^{n_1}(Y_i - \overline{Y})^2}}$$

is $t_{(n_1+n_2-2)}$ distributed.

Test of Hypotheses

We have already noted above, that quantities being functions of sample statistics and parameters may be applied for inference about these parameters. One such application is the testing of a hypothesis. We give a short illustration here for the situation of a single sample $X_1, \ldots, X_n$ from the normal distribution with unknown parameters μ and σ^2. But see e.g. [54, Chapter IX] for a more sound discussion.

Two-sided Hypothesis on the Expectation

Suppose we wish to decide whether the unknown expectation μ of our sample is equal to some given value μ_0 or not. Then this is the problem of testing the hypothesis $H_0 : \mu = \mu_0$ versus the alternative hypothesis $H_1 : \mu \neq \mu_0$. To come to a decision we may apply the test statistic

$$T = \sqrt{n}\,\frac{\overline{X} - \mu_0}{S}\,,$$

being in fact a statistic, since it does not contain unknown parameters. (The unknown μ in the quantity from Theorem 8.5 has been replaced by the hypothetical μ_0.) We already know that $\overline{X}$ is a reasonable estimator for the unknown μ, so that the test statistic measures the difference between a reasonable estimator for μ and the hypothetical value μ_0. Therefore, if the actual computed value of T turns out to be either too small or too large, then this indicates that the alternative hypothesis $H_1 : \mu \neq \mu_0$ is likely to be true.

Hence it remains to specify what 'too small' and 'too large' actually means. Since we know that the test statistic T is $t_{(n-1)}$ distributed, we may specify a small α (very often $\alpha = 0.05$) and say that the observed value of T is 'too small' if it is smaller than the $\alpha/2$ quantile $t_{(1-n),\alpha/2}$ of the $t_{(n-1)}$ distribution. We may then also say that the observation of T is 'too large' if it is greater than the $1 - \alpha/2$ quantile $t_{(1-n),1-\alpha/2}$ of the $t_{(n-1)}$ distribution. Hence, the hypothesis is rejected when

$$T < t_{(1-n),\alpha/2} \quad \text{or} \quad T > t_{(1-n),1-\alpha/2}\,.$$

Since the $t_{(\nu)}$ distribution is symmetric about 0, we have $t_{(1-n),1-\alpha/2} = -t_{(1-n),\alpha/2}$, and we may thus reject the hypothesis H_0 if $|T| > t_{(1-n),1-\alpha/2}$.

This procedure guarantees that if H_0 is in fact true, then the probability for rejecting H_0 is equal to α.

One-sided Hypothesis on the Expectation

Quite often, it is not necessary to decide whether μ equals a specific μ_0, but only if it is smaller than a specific μ_0 or not[8]. Then we test the hypothesis $H_0 : \mu \leq \mu_0$ versus the alternative hypothesis $H_1 : \mu > \mu_0$. We may apply the same test statistic as above, but now reject H_0 only if T is too large. We may specify T as being 'too large' if the actual value of T is greater than the $1 - \alpha$ quantile $t_{(1-n),1-\alpha}$ of the $t_{(n-1)}$ distribution. This guarantees that if H_0 is in fact true, the probability for rejecting H_0 cannot exceed α.

Hypothesis on the Variance

Similarly to the above, we may consider hypothesis of the form $H_0 : \sigma^2 = \sigma_0^2$ versus $H_1 : \sigma^2 \neq \sigma_0^2$ or $H_0 : \sigma^2 \leq \sigma_0^2$ versus $H_1 : \sigma^2 > \sigma_0^2$ for some given σ_0^2. Then the quantity

$$X^2 = (n-1)S^2/\sigma_0^2$$

is a statistic, having a chi-square distribution with $\nu = n-1$ degrees of freedom if $\sigma^2 = \sigma_0^2$. Based on this statistic, the test procedure is similar to the above procedure for testing a hypothesis about μ. The two-sided hypothesis is rejected when the observed value of X^2 is greater than $\chi^2_{(n-1),1-\alpha/2}$ or smaller than $\chi^2_{(n-1),\alpha/2}$. The one-sided hypothesis is rejected when the observed value of X^2 is greater than $\chi^2_{(n-1),1-\alpha}$.

8.4.3   Distribution of the Sample z-Score

Consider a sample $X_1, \ldots, X_n$ from the normal distribution with parameters μ and σ^2, where $n > 2$. Then the i-th sample z-score may be defined as

$$Z_i = \frac{X_i - \overline{X}}{D} \, ,$$

where $D^2 = \frac{1}{n} \sum_{i=1}^{n} (X_i - \overline{X})^2$. The observed sample z-score z_i can be seen as a measure of the relative standing of an individual observation x_i.

---

[8]We may of course also be interested in the problem whether the unknown μ is greater than μ_0, but the required adjustment of the described procedure is obvious.

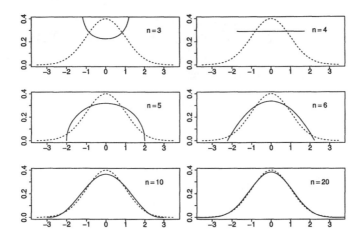

Fig. 8.9: Probability density function (solid line) of the sample z-score for different sample sizes n compared to the pdf of the standard normal distribution (dotted line)

What is the distribution of the random variable Z_i? First, it is obvious that the distribution of an individual Z_i does not depend on the position i (and is thus the same for each i), since the order of the random variables $X_1, \ldots, X_n$ is of no matter. For simplicity we will consider the n-th sample z-score Z_n in the following.

To derive the distribution of Z_n, we may consider the two independent samples $X_1, \ldots, X_{n-1}$ and X_n having the same expectation and variance, the former being of size $n_1 = n - 1$ and the latter being of size $n_2 = 1$. By letting $\overline{X}_* = \frac{1}{n-1} \sum_{i=1}^{n-1} X_i$, from Theorem 8.7 the statistic

$$T_* = \sqrt{\frac{n-1}{n}} \frac{(X_n - \overline{X}_*)}{\frac{1}{\sqrt{n-2}} \sqrt{\sum_{i=1}^{n-1} (X_i - \overline{X}_*)^2}}$$

has $t_{(n-2)}$ distribution. By using the identities

$$X_n - \overline{X} = \frac{n-1}{n} (X_n - \overline{X}_*)$$

and

$$\sum_{i=1}^{n}(X_i - \overline{X})^2 = \sum_{i=1}^{n-1}(X_i - \overline{X}_*)^2 + \frac{n-1}{n}(X_n - \overline{X}_*)^2$$

it follows that Z_n can be written as

$$Z_n = \frac{\sqrt{n-1}T_*}{\sqrt{T_*^2 + n - 2}} \, .$$

Since the distribution of T_* is $t_{(n-2)}$, the pdf of Z_n can be derived as

$$f(x) = \frac{1}{\sqrt{\pi(n-1)}} \frac{\Gamma\left(\frac{n-1}{2}\right)}{\Gamma\left(\frac{n-2}{2}\right)} \left(1 - \frac{x^2}{n-1}\right)^{(n-4)/2}$$

for

$$-\sqrt{n-1} < x < \sqrt{n-1} \, .$$

The corresponding distribution is known as the *Thompson distribution*[9] with parameter $\nu = n - 2$, see [87]. See also [16, p. 240/241] for a derivation of the pdf of the Thompson distribution with parameter $\nu = n - 1$. It can be shown that

$$\mathrm{E}(Z_n) = 0 \quad \text{and} \quad \mathrm{Var}(Z_n) = 1 \, .$$

For $n \to \infty$ the distribution of Z_n approaches the standard normal distribution, compare also Fig. 8.9.

Alternatively, one may define the i-th sample z-score as

$$\widetilde{Z}_i = \frac{X_i - \overline{X}}{S} = \sqrt{\frac{n-1}{n}} Z_i \, .$$

The distribution of $\widetilde{Z}_i$ may then easily be derived from that of Z_i.

---

[9]The pdf of the Thompson distribution is the same as that for the distribution of Z_n with n replaced by $\nu + 2$, where $\nu = 1, 2, \ldots$.

Bibliography

[1] D'Agostino, R.B. (1986): Tests for the normal distribution. In: D'Agostino, R.B. and Stephens, M.A., eds.: Goodness-of-Fit Techniques. Marcel Dekker, New York.

[2] Aitchison, J. and Brown, J.A.C. (1969): The Lognormal Distribution (with special reference to its uses in economics). Cambridge At The University Press, Cambridge.

[3] Akaike, H. (1974): A new look at statistical model identification. IEEE Transactions on Automatic Control, **AU-19**, 716–722.

[4] Anderson, T.W. and Darling, D.A. (1952): Asymptotic theory of certain 'goodness of fit' criteria based on stochastic processes. Annals of Mathematical Statistics, **23**, 193–212.

[5] Atkinson, A.C. (1985): Plots, Transformations, and Regression. Clarendon Press, Oxford.

[6] Azzalini, A. (1985): A class of distributions which includes the normal ones. Scandinavian Journal of Statistics, **12**, 171–178.

[7] Balanda, K.P. and MacGillivray, H.L. (1988): Kurtosis: a critical review. American Statistician, **42**, 111–119.

[8] Blom, G. (1958): Statistical Estimates and Transformed Beta Variables. Wiley, New York.

[9] Box, G.E.P. and Muller, M.E. (1958): A note on the generation of random normal deviates. Annals of Mathematical Statistics, **29**, 610–611.

[10] Box, G.E.P. and Cox, D.R. (1964): An analysis of transformations (with discussion). Journal of the Royal Statistical Society, Ser. B, **26**, 211–246.

[11] Broffitt, J.D. (1986): Zero correlation, independence, and normality. The American Statistician, **40**, 276–277.

[12] Casella, G. and Berger, R.L. (2002): Statistical Inference. Second Edition. Duxbury, Australia.

[13] Chou, Y.-M. (1981): Additions to the table of normal integrals. Communications in Statistics, Ser. B, **10**, 537–538.

[14] Chou, Y.-M., Turner, S., Henson, S., Meyer, D. and Chen, K.S. (1994): On using percentiles to fit data by a Johnson distribution. Communications in Statistics, Ser. B, **23**, 341–354.

[15] Cook, R.D. and Weisberg, S. (1994): An Introduction to Regression Graphics. Wiley, New York.

[16] Cramér, H. (1971): Mathematical Methods of Statistics (12. print.). Princeton University Press, Princeton.

[17] Crow, E.L. and Shimizu, K., eds., (1988): Lognormal Distributions: Theory and Applications. Dekker, New York.

[18] Dalgaard, P. (2002): Introductory Statistics with R. Springer, New York.

[19] Dallal, G.E. and Wilkinson, L. (1986): An analytic approximation to the distribution of Lilliefors' test for normality. The American Statistician, **40**, 294–296.

[20] Day, N.E. (1969): Estimating the components of a mixture of normal distributions. Biometrika, **56**, 463–474.

[21] Dempster, A.P., Laird, N.M. and Rubin, D.B. (1977): Maximum likelihood from incomplete data via the EM algorithm. Journal of the Royal Statistical Society, Ser. B., **39**, 1–38.

[22] Fisher, R.A. (1922): On the interpretation of χ^2 from contingency tables and calculation of P. Journal of the Royal Statistical Society, Ser. A, **85**, 87–94.

[23] Fisher, R.A. (1924): On a Distribution Yielding the Error Functions of Several well Known Statistics. In: Proceedings of the International Mathematics Congress, Toronto.

[24] Freedman, D. and Diaconis, P. (1981): On the histogram as a density estimator: L_2 theory. Zeitschrift für Wahrscheinlichkeitstheorie und verwandte Gebiete, **57**, 453–476.

[25] Galton, F. (1889): Natural Inheritance. Macmillian, London.

[26] Gaudard, M. and Karson, M. (2000): On estimating the Box-Cox transformation to normality. Communications in Statistics, Ser. B. , **29**, 559–582.

[27] Gauss, C.F. (1809): Theoria Motus Corporum Coelestium in Sectionibus Coincis Solem Ambientium. F. Perthes and I.H. Besser, Hamburg.

[28] Groß, J. (2003): Linear Regression. Springer, Berlin.

[29] Guerrero, V.M. and Johnson, R.A. (1982): Use of the Box-Cox transformation with binary response models. Biometrika, **69**, 309–314.

[30] Harter, H.L. (1961): Expected values of normal order statistics. Biometrika, **48**, 151–165. Correction: p. 476.

[31] Hastings, C. (1955): Approximations for Digital Computers. Princeton University Press, Princeton.

[32] Hegde, L.M. and Dahiya, R.C. (1989): Estimation of the parameters in a truncated normal distribution. Communications in Statistics, Ser. A, **18**, 4177–4195.

[33] Henze, N. (1986): A probabilistic representation of the 'skew-normal' distribution. Scandinavian Journal of Statistics, **13**, 271–275.

[34] Hosmer, D.W. (1973): On MLE of the parameters of a mixture of two normal distributions when the sample size is small. Communications in Statistics, Theory and Methods, **10**, 217–227.

[35] Johnson, N.L. (1949): Systems of frequency curves generated by methods of translation. Biometrika, **36**, 149–176.

[36] Johnson, N.L., Kotz, S. and Balakrishnan, N. (1994): Continuous Univariate Distributions. Volume 1. Second Edition. Wiley, New York.

[37] Johnson, N.L., Kotz, S. and Balakrishnan, N. (1995): Continuous Univariate Distributions. Volume 2. Second Edition. Wiley, New York.

[38] Kuiper, N.H. (1960): Tests concerning random points on a circle. Proceedings, Akademie van Wetenschappen, **A 63**, 38–47.

[39] Kunert, J., Montag, A. and Pöhlmann, S. (2001): The quincunx: history and mathematics. Statistical Papers, **42**, 143–169.

[40] Lancaster, H.O. (1959): Zero correlation and independence. Australian Journal of Statistics, **21**, 53–56.

[41] Laplace, P.S. (1774): Determiner le milieu que l'on doit prendre entre trois observations données d'un même phénomené. Mémoires de Mathématique et Physique presentées à l'Académie Royale des Sciences par divers Savans, **6**, 621–625.

[42] Leone, F.C., Nelson, L.S. and Nottingham, R.B. (1961): The folded normal distribution. Technometrics, **3**, 543–550. [Corrigenda: **22**, 452.]

[43] Lilliefors, H.W. (1967): On the Kolmogorov-Smirnov test for normality with mean and variance unknown. Journal of the American Statistical Association, **62**, 399–402.

[44] Lehmann, E.L. and Casella, G. (1995): Theory of Point Estimation. Second Edition. Springer, New York.

[45] Maindonald, J. and Brown, J. (2003). Data Analysis and Graphics Using R. Cambridge University Press, Cambridge.

[46] Marden, I. (1998): Bivariate QQ-plots and spider web plots. Statistica Sinica, **8**, 813–826.

[47] Mardia, K.V., Kent, J.T. and Bibby, J.M. (1979): Multivariate Analysis. Academic Press, London.

214   BIBLIOGRAPHY

[48] Mellnick, E.L. and Tenenbein, A. (1987): Comment on Broffitt (1986). The American Statistician, **41**, 243.

[49] Mellnick, E.L. and Tenenbein, A. (1982): Misspecifications of the normal distribution. The American Statistician, **36**, 372–373.

[50] Mittal, M.M. and Dahiya, R.C. (1987): Estimating the parameters of a doubly truncated normal distribution. Communications in Statistics, Ser. B, **16** 141–159.

[51] McClave, J.T., Benson, P.G. and Sinich, T. (2001). Statistics for Business and Economics. Eigth Edition. Prentice Hall, Upper Saddle River.

[52] de Moivre, A. (1733): Approximatio ad Summam Ferminorum Binomii $(a + b)^n$ in seriem expansi. Supplementum II Miscellanae Analytica, 1–7.

[53] de Moivre, A. (1756): The doctrine of chance. Third Edition. [Reprinted 1967. Chelsea, New York.]

[54] Mood, A.M., Graybill, F.A. and Boes, D.C. (1974): Introduction to the Theory of Statistics. Third Edition. McGraw-Hill, Auckland.

[55] Moore, D.S. (1986): Tests of the chi-squared type. In: D'Agostino, R..B. and Stephens, M.A., eds.: Goodness-of-Fit Techniques. Marcel Dekker, New York.

[56] Moran, P.A.P. (1980): Calculation of the normal distribution function. Biometrika, **67**, 675–676.

[57] Mosteller, F. and Tukey, J.W. (1977): Data Analysis and Regression. Addison-Wesley, Reading.

[58] Owen, D.B. (1980): A table of normal integrals. Communications in Statistics, Ser. B, **9**, 389–419. [Corrigenda: **10**, 541.]

[59] Owen, D.B. (1988): The starship. Communications in Statistics, Ser. B, **17**, 315–323.

[60] Owen, D.B. and Li, H. (1988): The starship for point estimates and confidence intervals on a mean and for percentiles. Communications in Statistics, Ser. B, **17**, 325–341.

[61] Parrish, R.S. (1992): Computing expected values of normal order statistics. Communications in Statistics, Ser. B, **21**, 57–70.

[62] Parrish, R.S. (1992): Computing variances and covariances of normal order statistics. Communications in Statistics, Ser. B, **21**, 71–101.

[63] Patel, J.K. and Read, C.B. (1996): Handbook of the Normal Distribution. Second Edition, Revised and Expanded. Marcel Dekker, New York.

[64] Pearson, K. (1894): Contributions to the mathematical study of evolution. Philosophical Transactions of the Royal Society of London, Ser. A, **185**, 71–110.

[65] Pearson, K. (1900): On the criterion that a given system of deviations from the probable in the case of a correlated system of variables is such

that it can be reasonably supposed to have arisen from random sampling. Philosophical Magazine, **50**, 157–175.

[66] Pearson, K. (1924): Historical note on the origin of the normal curve of errors. Biometrika, **16**, 402–404.

[67] Rao, C.R. (1973): Linear Statistical Inference and Its Applications. Second Edition. Wiley, New York.

[68] Robert, C. (1991): Generalized inverse normal distributions. Statistics & Probability Letters, **11**, 37–41.

[69] Romano, J.P. and Siegel, A.F. (1986): Counterexamples in Probability and Statistics. Wadsworth, Belmont California.

[70] Ross, S.M. (1997): Simulation. Second edition. Academic Press, San Diego.

[71] Royston, P. (1982): An extension of Shapiro and Wilk's W test for normality to large samples. Applied Statistics, **31**, 115-124.

[72] Royston, P. (1993): A pocket-calculator algorithm for the Shapiro-Francia test for non-normality: an application to medicine. Statistics in Medicine, **12**, 181–184.

[73] Schneider, H. (1986): Truncated and Censored Samples from Normal Populations. New York, Marcel Dekker.

[74] Scott, D.W. (1979): On optimal and data-based histograms. Biometrika, **66**, 605–610.

[75] Shapiro, S.S. and Wilk, M.B. (1965): An analysis of variance test for normality (complete samples). Biometrika, **52**, 591–611.

[76] Shapiro, S.S. and Francia, R.S. (1972): Approximate analysis of variance test for normality. Journal of the American Statistical Association, **67**, 215–216.

[77] Slifker, J.F and Shapiro, S.S. (1980): The Johnson system: selection and parameter estimation. Technometrics, **22**, 239–246.

[78] Snedecor, G.W. (1934): Calculation and Interpretation of Analysis of Variance and Covariance. Collegiate Press, Ames, IA.

[79] Snedecor, G.W. and Cochran, W.G. (1967): Statistical Methods. Sixth Edition. Iowa State University Press, Ames, Iowa.

[80] Stephens, M.A. (1974): EDF statistics for goodness of fit and some comparisons. Journal of the American Statistical Association, **69**, 730–737.

[81] Stephens, M.A. (1986): Tests based on EDF statistics. In: D'Agostino, R.B. and Stephens, M.A., eds.: Goodness-of-Fit Techniques. Marcel Dekker, New York.

[82] Stephens, M.A. (1986): Tests based on regression and correlation. In: D'Agostino, R.B. and Stephens, M.A., eds.: Goodness-of-Fit Techniques. Marcel Dekker, New York.

[83] Stuart, A. and Ord, J.K. (1994): Kendall's Advanced Theory of Statistics. Volume 1. Distribution Theory. Sixth Edition. Edward Arnold, London.

[84] Student (Gosset, W.S.) (1908): The probable error of a mean. Biometrika, **6**, 1–25.

[85] Swift, A. L. (1994): Experience with a criterion for the classification of Johnson distributions. Communications in Statistics, Ser. B, **23**, 1127–1136.

[86] Thode Jr., H.C. (2002): Testing for Normality. Marcel Dekker, New York.

[87] Thompson, W.R. (1935): On a criterion for the rejection of observations and the distribution of the ratio of deviation to sample standard deviation. The Annals of Mathematical Statistics, **6**, 214–219.

[88] Tong, Y.L. (1990): The Multivariate Normal Distribution. Springer, New York.

[89] Venables, W.N., Ripley, B.D. (2002): Modern Applied Statistics with S. Fourth Edition. Springer, New York.

[90] Wheeler, R.E. (1980): Quantile estimators of Johnson curve parameters. Biometrika, **67**, 725–728.

[91] Wilson, E.B. and Hilferty, M.M. (1931): The distribution of chi-square. Proceedings of the National Academy of Sciences, **17**, 684–688.

Index

Daniela Lohlein

An Economic Analysis of Public Good Provision in Rural Russia

The Case of Education and Health Care

Frankfurt am Main, Berlin, Bern, Bruxelles, New York, Oxford, Wien, 2003.
XIV, 165 pp., 29 fig., 53 tab.
Development Economics and Policy.
Edited by Franz Heidhues and Joachim von Braun. Vol. 36
ISBN 3-631-50774-7 / US-ISBN 0-8204-6427-9 · pb. € 39.–*

Adequate access to public goods plays an important role in the economic and rural development of any country. For the transition countries, however, providing public goods is particularly problematic owing to several factors associated with the transition process itself, e.g. economic recession and a change in ownership of local public goods. Taking Russia as an example, this study examines the effect of the transition process on rural households' access to public goods. With reference to education and health care, household access to public goods is addressed in terms of community availability and economic access. The analysis is taken a step further, through an examination of the role of informal institutions in public good provision. Multiple regression analysis was used to test for the significance of income as a determinant of private expenditures on public goods. The results indicate that, contrary to expectations, neither income nor informal payments are important determinants of access to public goods. Informal institutions continue to exert a strong influence on the provision of public goods.

Contents: Public good provision in transition countries · Decentralisation and public good provision · Education and health care provision in rural Russia during transition · Household access to education and health care · Institutional analysis of local public good provision · Role of informal institutions in local public good provision · Household and community survey · Regression analysis · New Institutional Economics

Frankfurt am Main · Berlin · Bern · Bruxelles · New York · Oxford · Wien
Distribution: Verlag Peter Lang AG
Moosstr. 1, CH-2542 Pieterlen
Telefax 00 41 (0) 32 / 376 17 27

*The €-price includes German tax rate
Prices are subject to change without notice
Homepage http://www.peterlang.de